T0256724

The Political Economy of Pipelines

The Political Economy of Pipelines

A Century of Comparative Institutional Development

JEFF D. MAKHOLM

THE UNIVERSITY OF CHICAGO PRESS   CHICAGO AND LONDON

JEFF D. MAKHOLM is senior vice president at NERA Economic Consulting, an international firm providing economic analysis to corporations, governments, and regulatory
agencies. He is the author of *The Distribution and Pricing of Sichuan Natural Gas.*

The University of Chicago Press, Chicago 60637
The University of Chicago Press, Ltd., London
© 2012 by The University of Chicago
All rights reserved. Published 2012.
Printed in the United States of America
21 20 19 18 17 16 15 14 13 12   1 2 3 4 5

ISBN-13: 978-0-226-50210-6 (cloth)
ISBN-10: 0-226-50210-4 (cloth)

Library of Congress Cataloging-in-Publication Data
Makholm, Jeff D.
    The political economy of pipelines : a century of comparative institutional development /
Jeff D. Makholm.
        pages ; cm.
    Includes bibliographical references and index.
    ISBN-13: 978-0-226-50210-6 (cloth : alkaline paper)
    ISBN-10: 0-226-50210-4 (cloth : alkaline paper)   1. Petroleum pipeline industry—
Economic aspects.   2. Petroleum pipeline industry—Government policy—United
States.   3. Petroleum industry and trade—Economic aspects.   4. Petroleum law and
legislation—United States.   5. Institutional economics.   I. Title.
    HD9580.A2M354 2012
    388.5'50973—dc23   2011029351

♾ This paper meets the requirements of ANSI/NISO Z39.48-1992 (Permanence of Paper).

TO LAUREN, KARLA, AND DANIEL

Contents

Preface

This book uses a century of the world's experience with oil and gas pipelines to illustrate the importance of the new institutional economics in explaining markets, market behavior, regulation, and competitive entry. The industry is an excellent illustration of these theories because of its great institutional diversity in the face of a remarkable commonality in technology and operation no matter where pipelines are. Why are oil pipelines in the United States and gas pipelines in Europe vertically integrated while their close-cousin US gas pipelines are vertically separate? Why do gas pipelines in Victoria sell pipeline capacity in a regulated spot market while those in neighboring New South Wales sell through long-term contracts? Why do shippers on gas pipelines in the United States enjoy capacity as a property right (to buy and sell as they wish for market prices) while shippers in Canada do not? Why is new gas pipeline licensing in the United States a snap while the same in Europe constitutes high political drama? Why does credit and investor capital flow easily to US gas pipelines (the world's oldest) while those in Argentina (the second oldest) have no credit? Ultimately, why is gas pipeline use and new construction a genuinely competitive activity in the United States but nowhere else in the world? Traditional neoclassical economic analysis, with its focus on technology and costs, rational choice, and equilibrium, has little to say about why pipelines play such different roles in their nations' and regions' economies. The new institutional economics, drawing upon a wider theoretical base, is more sensitive to thick and local knowledge and the institutional and political histories that have shaped—and continue to shape—pipeline transport markets around the world.

The pipeline industry is inherently difficult to study. While all oil and gas pipelines perform similar physical functions, the institutions that surround them are stubbornly dissimilar from country to country (and even in oil versus gas). The institutions that shaped the development of oil and gas pipelines in the United States arose more than a century ago in response to public disgust over industry practices that are now largely forgotten. The institutions surrounding the great investor-owned pipeline systems in Europe, Australia, or South America are rather new by comparison, as they developed with the pipeline privatizations of the late twentieth century. But even the latter examples are quite different from region to region, reflecting idiosyncratic local issues, market structures, and politics. The institutions and experiences from one region do not translate to others, and those differences are more often than not buried in documents and experiences that are difficult for outsiders to understand.

My work at National Economic Research Associates (NERA) has been closely connected with pipelines and the market and regulatory conflicts that surround them. When the firm was founded in 1961, its leaders already represented gas distributors in one of the great conflicts arising from the way Congress created legislation for US gas pipelines in the 1930s. Experience with that and related conflicts led to my own direct participation in two major innovations involving the world's pipeline businesses. The first was the creation of the market in legal entitlements for gas transport—the "Coasian" market—in the United States. The second came with the worldwide privatization wave of the late 1980s and 1990s, when I traveled around the globe to help to create and/or regulate investor-owned pipeline companies out of government-owned gas pipeline departments. In the process, I worked directly with most of the world's major gas and oil pipeline systems, in North and South America, Europe, Australia, New Zealand, Russia, and China.

During this period, I encountered the diversity in the social and economic organization of pipeline systems. My work included setting prices for the communist-era pipelines in Poland, working extensively with British Gas (the first in the late twentieth-century wave of pipeline privatizations), helping to privatize pipeline systems in Argentina and Bolivia, and recommending reforms for the state-owned gas pipeline system in China. I was involved in the privatization and regulation of the gas pipeline systems in Australia and New Zealand, in creating tariff systems for the state-owned oil and gas pipeline systems in Mexico and South Africa, and in the investigation of whether the oil pipeline system in Russia might be a candidate for Western investor capital. I also advised on how to finance new gas pipeline development in regions with unconvertible currencies and little credit (e.g., East and West Africa).

I analyze in this book how economic theory captures the development and regulation of pipelines. Not every major pipeline system provides a useful illustration of economic principles. Social conflict between private enterprise and the public welfare does not occur for state-owned pipelines, such as those in Russia and China, as it does for investor-owned pipelines. State-owned pipelines are thus not part of my analysis. In order to keep this book to a manageable length, I have omitted discussion of the sources and techniques for making regulated pipeline tariffs. Much of my work has involved the creation of pipeline tariff regimes for privatized pipelines. I believe that pipeline tariff setting should be uncomplicated and straightforward as long as the definition of regulated pipeline property is unambiguous and the operation and financing of pipeline systems are transparent to public scrutiny. Unfortunately, clear property rights are often lacking—and genuine transparency is rare— for the world's newly privatized pipelines. I have also omitted pipeline maps, other than to provide examples, from sources now very hard to obtain, of what legislators saw when they first regulated pipelines in the United States. What pipeline systems look like on a map is rarely important in performing economic analysis—the US oil and gas pipeline systems look very similar in this way but have almost nothing in common in terms of pricing, transacting, and regulation. In any event, pipeline maps are widely available on the web for anyone who cares to look.

I owe a great intellectual debt to my senior colleagues, including Louis Guth, Michael Tennican, Bernard Reddy, Graham Shuttleworth, William Taylor, Agustin Ros, and Greg Houston, who all read parts of the manuscript and made valuable suggestions. Mario Lopez read the entire manuscript and provided very helpful connections to the most recent theoretical work in transaction cost economics and related fields. I also received gratefully the comments of Fernando Vinelli, Paul Hunt, and Christian von Hirschhausen. Donald Kaplan was kind enough to share otherwise unobtainable records from his time at the US Department of Justice. Wayne Olson provided many valuable suggestions. Assisting substantially with my research were Kurt Strunk, Gabriel Prieto, Joshua Rogers, Ryan Knight, and Alexander Walsh. Simone Cote has been a sensitive editor of my sometimes tangled prose. Two anonymous reviewers of the early manuscript had a profound influence on this book's ultimate focus. Such contributors remain the unsung heroes in scholarly publishing, and I regret that I cannot thank them by name— nor, according to the policy of the University of Chicago Press, can I thank my editor here. Anyhow, whatever mistakes or oversights remain are mine alone.

This book reflects the influence of two economists in particular. One is the late Leonard Weiss, professor of economics and for three decades the leader in the study of industrial organization at the University of Wisconsin–Madison. Weiss is widely credited with the first collection of industry case studies that blended economic theory, historical evidence, and policy. He was my bridge to a long tradition in Wisconsin's Department of Economics, stretching back to the institutional economist John R. Commons. That tradition holds that the pragmatic study of the intersection of institutions and economic theory produces the firmest foundation for economic and regulatory policy in the public interest. I hope that that Wisconsin tradition is alive and well in this book. The other is Alfred E. Kahn, professor, dean, and trustee at Cornell University; chairman of the New York Public Service Commission and the Civil Aeronautics Board; deregulator of airlines; presidential adviser; and my longtime friend and colleague at NERA—just recently deceased. Kahn was a direct witness to the people and events that have shaped pipeline policies since the 1940s and was a kind and insightful—not to mention vigorous—reviewer of my manuscript. He will always be for me, as for the many economists who worked with him in studying and shaping economic policy for more than seven decades, the archetype of boundless inspiration, wisdom, grace, and good humor.

The New Institutional Economics and Pipeline Transport

W hen President Jimmy Carter appointed Professor Alfred E. Kahn, of Cornell University, to reform the US airline industry in the late 1970s, Kahn described the nature of that business as *marginal costs with wings*—capable of moving its highly deployable capital as its market demanded, as long as it was unhindered by complex entry, exit, and pricing regulation.[1] The subsequent abolition of the Civil Aeronautics Board (the US airline rate-setting authority) was, however, a more complicated affair than simply turning air carriers loose to compete. The development of hub-and-spoke routing and lack of capacity at some major hubs showed that the consequences of airline deregulation were unpredictable and more complicated than many foresaw.[2] But in the end, airline capital did prove to be magnificently mobile—marginal costs with wings. The competitive response by carriers to deregulation, both in pricing flexibility and in routing—to say nothing of the demise of inefficient carriers—revolutionized air travel in the United States, bringing whole new generations of discretionary travelers into the air for quicker and safer travel. Lifting price regulation and abolishing the regulatory agency was a triumph of neoclassical economics over the protectionist forces that had used regulation to support the cartelization of a structurally competitive airline transport industry.

This book is a story about the triumph of latter-day economic institutionalists—rather than neoclassical economists—in providing the theoretical perspective to analyze how and why competition arose in the pipeline transport industry. So why begin by recalling airline deregulation? It provides such a usefully sharp contrast. Pipelines are *marginal costs with a ball and chain.* Pipeline capital is land bound and immobile—the antithesis of deployable capital in air transport. Pipelines are built from one spot to a distant one. By far the most efficient method of inland fuel transport, pipelines serve particular oil and gas producers at one end and refineries, gas distributors, or power plants at the other—often a continent away. Uncertainty or commercial opportunism at either end of the pipe, by the pipelines or their users, can strand facilities and wreck the value of the invested capital. The challenges are so great that, often enough, governments are left to build the pipelines themselves with public funds. If investors build pipelines, they make interlocking alliances with fuel suppliers and users. Promoting competitive pipeline transport, in the face of the immobility of capital and those resulting alliances, involves more complex problems than those Kahn overcame in promoting competition in air transport. Explaining the source of competitive pipeline transport calls for a more diverse economic theory.

The interlocking alliances that the pipeline industry requires lie at the heart of the matter. Chief among them is vertical integration. But vertical integration is problematic: oil and gas producers tend to use vertically integrated pipelines as weapons against nonintegrated rivals. Governments have taken various approaches to facilitating the flow of capital to pipeline transport—to efficiently serve the public's need for fuel—while trying to prevent oil- and gas-producing companies from using pipeline access to foil competition in fuel markets. The United States and Canada sought a solution through the regulation of investor-owned pipelines by appointed commissions, with widely varying measures of success over the course of the twentieth century. Most of the rest of the world turned to government-owned pipelines and various forms of direct state control. But the choice of private or public capital has had profound consequences. In North America the use of private pipeline capital helped shape the development—decades ago—of regulatory institutions of the sort required to pave the way for competitive pipeline transport. Why did North America, almost alone, choose that path?

What about competition itself? The signal qualities of pipelines might seem to bode poorly for competitive inland transport: long-lived, stationary pipelines and corresponding long-term agreements needed to protect capital investments. Indeed, the investor-owned pipeline industry's first century was typified not by competitive transport but by a combination of vertical integration and heavy regulation in combination with regulated fuel markets. In the twenty-first century, however, while the investor-owned gas pipeline system in the United States retains a regime of comprehensive cost-based regulated tariffs developed decades ago, it simultaneously exhibits unregulated—and highly competitive—markets for both the use and expansion of the pipeline system. Those competitive transport markets in turn support highly competitive spot and futures gas commodity markets. How did such competitive pipeline transport develop? Why did it take a century to appear? And why has the success of competitive gas pipeline transport in the United States not spread to other pipeline systems—even to that gas pipeline system's close oil pipeline cousin in the United States?

These are puzzling questions for economists. Pipelines are pipelines, in technology and operation, no matter where they are. But economic theories grounded in profit or welfare maximization, rational choice, and equilibrium—the foundations of modern neoclassical economics, which has dominated the field since 1941[3]—provide no foothold for dealing with such questions. To be sure, most modern economists owe a debt to the neoclassical economic tradition for the order, rationality, and mathematical logic that it provided, as it displaced the essentially non-mathematical institutional economics of the early twentieth century. But while a neoclassical perspective embraces some elements of the basic cost structure of pipelines, it is not a sufficient analytical tool to address the organization of the pipeline industry. Finding common threads in the economic analysis of the industry means embracing the more diverse, interdisciplinary theoretical perspectives of what has been called the new institutional economics—relatively recent extensions and developments of economic theory.[4] The institutional details and regional histories shaping complex industries—particularly transport—are important for those latter-day institutionalists who have developed the new institutional economics.[5] Such theories help to explain the choice between employing private or public capital for pipeline projects; why the industry structure in one region diverges sharply from those in others; how contracts in some regions supplanted vertical integration; and why competitive transport has yet happened on only one pipeline system in a world of pipeline systems.

The new institutional economics employs such diverse analytical perspectives as legal institutions in a market economy, modes of industrial
governance, contractual arrangements, public (political) choice, regulation, and institutional change.[6] Four of its elements go far to explain the
organizational diversity exhibited in the world's pipeline industry: transaction costs and asset specificity, institutional evolution, intangible property rights, and collective action.

Transaction Costs and Asset Specificity

Asset specificity as a concept arose among economists to explain why
firms vertically integrate rather than either contract or deal in spot markets with one another. Certain kinds of investments are so sunk and dedicated to particular business relationships that they give rise to the risk
of opportunistic "holdup"—a problem that vertical integration serves to
allay. Pipelines display great asset specificity: immobile assets of great
length tied to fuel producers, oil refiners, power plants, or local gas distributors. Vertical integration ties the interests of those producers, pipeline companies, refineries, power plants, or gas distributors together.
Such ties greatly lessen—if not eliminate—the prospect that one party
in such fuel transport arrangements can hold up another and limit the
expected return on the sunk capital investments involved. The problem
with pipelines is that while asset specificity pulls pipelines to vertically
integrate, their inherent economies of scale limit their number, thus con-
centrating fuel markets around a relatively small number of vertically integrated pipeline companies.

Oliver Williamson, of the University of California–Berkeley, shared
the Nobel Prize in Economics in 2009 for his theoretical work in *transaction cost economics* that deals with just such attributes. The theory
explains why some economic transactions take place inside firms and
others happen by contract in the marketplace. He coined the term *asset
specificity* around 1980. Asset specificity compels investors to forge reliable commercial relationships *before* building pipelines. Building first
and negotiating later would allow customers to take advantage of the
immobility of investors' committed capital to extract concessions and
sharply limit profitability.

Dealing with pipelines' asset specificity requires that producers, pipeline companies, refiners, gas distributors, and others transact reliably with one another. Such transacting has costs of two sorts: the lesser involves ex ante negotiating and drafting costs; the greater involves the ex post costs that arise if joint agreements fail to survive unforeseen events.[7] A central tenet of transaction cost economics is that the contracts that define such relationships are always necessarily incomplete—the result of what Williamson called "bounded rationality."[8] While human actors have the capacity to look ahead, uncover contractual hazards, and plan contractual and institutional arrangements accordingly, they can never eliminate such costs. Vertical integration, which avoids the potentially heavy ex post contracting costs, may seem to be the sine qua non for major oil and gas pipelines.

Neoclassical economic theory has a tendency to assume away transaction costs.[9] In a world without transaction costs, decision makers possess perfect foresight. They can effortlessly write complete, uncontroversial, binding contracts. In such a world, economic governance institutions play a neutral role in the efficiency of the productive process. It does not matter whether production is organized via prices in spot markets or within a vertically integrated firm.[10] Such perspectives cannot help but impair the analysis of industries for which such costs are so important. With so much riding on the ability of pipelines and their users to avoid the potentially costly consequences of asset specificity, Williamson's theoretical work on the costs and uncertainties of transacting rises to a level of preeminent importance.

That firms in general are specialized governance structures built to deal with the cost of transacting was proposed by economists long ago.[11] Ronald Coase's famous 1937 paper on the nature of the firm focused on the choice between contracting with suppliers and integrating with them.[12] Coase (the 1991 Nobel laureate in economics), of the London School of Economics and later the University of Chicago, had the insight that there are costs to using what he called the *price mechanism* (his term for spot markets) manifested by transaction, coordination, and contracting costs.[13] Williamson extended Coase's insight by examining not just the cost of contracting but the threat of opportunistic behavior by contracting parties that can be forestalled through vertical integration.[14]

For investor-owned pipelines, the costs of contracting dominated early relationships and led inexorably to vertical integration. The advent of the pipeline industry occurred during the early years of the US

Industrial Revolution. Those first Standard Oil pipelines appeared before the era of timely and reliable business records, legislated regulatory accounting rules, reliable regulatory administrative procedures, or the Securities and Exchange Commission. The idea did occur to some of those who debated the first pipeline regulations early in the twentieth century that pipelines might merely serve as independent long-distance transport companies, like railroads or canals. But sharper legislative minds saw that these latter two transport systems embodied versatility in their respective markets that pipelines did not. Canals and railroads could develop as their diverse economic product bases developed without the necessity of vertical integration. Those same sharp minds of a century ago also saw that regulating pipelines without regard to their need to transact reliably at either end could doom the prospect for private pipeline funding or harm independent fuel producers or both.[15]

The use of private investor capital for early US oil and gas pipelines put a spotlight on the inability to use then-existing institutions to deal effectively with the cost of transacting through contracts. The consequence, for both oil and gas, was the concentration of the industries around a limited number of vertically integrated pipelines by the 1930s. The options for dealing with the problem were few for oil pipelines with their existing common carriage legislation that explicitly forbade contracts. Gas pipeline regulation presented a blank slate, however, and the congressional response was the development of legal and accounting institutions that would lower the cost of pipeline contracting and facilitate the move toward an independent pipeline industry capable of attracting private capital.

Indeed, an array of institutions combined to bring down the cost of transacting and clear a path for the evolution of the pipeline industry away from the seeming sine qua non of vertical integration toward an independent and competitive inland transport business. Some of the basic institutions predated pipelines; others accompanied the early twentieth-century efforts in the United States to regulate investor-owned public utilities effectively; still others dealt with pipelines directly to remedy abusive or acquisitive practices. But such institutions do not generally appear outside of North America. If and when such institutions arose to reduce contracting costs explains much of the geographic and sector-specific (i.e., oil or gas) diversity exhibited by the various major pipeline systems in the twenty-first century. Economic theory that abstracts from the cost of transacting sees none of this.

Institutional Evolution and Market Development

The pipeline industry is old, as are some of the institutions that govern its behavior. As a specialized form of highly capital-intensive, long-distance inland transportation, pipelines descend from the older canal and railroad systems that developed in Europe and America to transport commodities during the first half of the nineteenth century. The roots of regulated inland transportation stretch back even further to how kings granted franchises and limited entry for carriers—thereby ensuring profitability—in exchange for commitments to serve all comers (i.e., their subjects). The governance institutions surrounding pipelines are thus complex products of relatively ancient social custom, public opinion, legislative action, and judicial precedent. As in other spheres of democratic governance, such institutions evolve. And more often than not, the evolution looks less like Charles Darwin's "gradualism" and more like the late evolutionary biologist Stephen J. Gould's "punctuated equilibrium"—with episodic evolutionary leaps resulting from current events often now long forgotten. Even then, such evolution is generally selective and incomplete. In the world of public choice and political evolution, not everything is up for grabs all the time.[16] Regulation itself is slow to evolve: with administrative and regulatory institutions led by practical and politically minded people, *what works* is hard to wipe away with new methods and new reasoning. A useful economic analysis of the diverse institutions guiding pipeline industry behavior must recognize their often-piecemeal development.

Douglass North (an economic historian who shared the Nobel Prize in Economics in 1993), of Washington University in Saint Louis, gained recognition for his studies of how and why such institutions of economic governance evolve to prompt economic growth. His criticism of neoclassical economic theory was that it ignored both those governance institutions and the factor of time in explaining the source of economic growth. North used the developmental history of canals, railroads, and ocean shipping to illustrate the extent to which his theory—the evolution of economic governance institutions in the pursuit of profit—drove advances in transport systems and the ready availability of investor capital to finance them. That he used transport to illustrate his insights is perhaps no surprise. Long-distance transport embodies a unique collection of problems in technology, finance and credit, property rights, public access, land use, and uncertainty. Advances in transport are central to the success of economies and depend utterly on reliable institutional foundations. It is to be expected that whatever collection of institutions evolved to handle transport are themselves complex.

North's theoretical contributions defy easy simplification.[17] But his historical examples are compelling. He illustrated that productivity and growth in ocean shipping in the early nineteenth century rose sharply not because of new technology but because of new institutions governing ocean trade.[18] He also demonstrated how growth in agricultural productivity in the United States early in the nineteenth century took place because new institutions for capital financing and safeguarding property rights made large-scale canal and rail transport systems possible. In such examples, particular interest groups (e.g., rail shippers, land developers, and farmers) learned how to use governing institutions to facilitate the raising of private investment funds, to enforce trading rules, and to protect private property rights. Sometimes these new institutions failed, such as public financing for transport in the United States when the state of New York defaulted on its Erie Canal bonds during the depression of 1839–42 rather than raise state taxes.[19] Other institutions, such as the Interstate Commerce Act and other legislation aimed at railroads, crushed some forms of reasonable price discrimination and did more to hasten the development of interregional railroad cartels than to promote efficient or competitive rail transport.[20] North's innovation was the deep study of the history of those transport modes to discern how interest groups pressed for (or took advantage of) new governing institutions to surmount obstacles to development and profitability.

One of North's criticisms of traditional neoclassical economic theory was its presumption that changes in technology and the use of new factors of production contributed to economic growth. He argued that new technology and better factor utilization are not the cause of growth but rather *constitute* growth, and that underlying institutional change is the root cause. Railroad and canal systems usefully illustrate his theory. Pipelines provide better examples still, perhaps, having none of the product versatility of those older transport modes. And unlike canals (whose role was superseded by railroads) and railroads (whose role was largely diminished in the twentieth century by other transport modes), pipelines have remained for more than a century the preferred method for long-distance inland energy transport—with a technology that has changed only unremarkably since steam shovels replaced hand shovels and welds replaced rivets and screw fittings. What better an industry than pipelines, with their single purpose and static technology, to confirm that institutional change pulled competitive rivalry out of entry-deterring monopoly or that disparate institutional foundations account for the current divide between European and North American competitiveness?

A particular focus of North's was identifying the creation of institutions to define and ensure property rights as the basis for new forms of financing and trade. The scrupulous definition of private property rights in pipeline transport similarly has been the critical factor in the development of an independent and competitive pipeline transport sector. Indeed, it was trouble with private pipeline capital in the United States that led legislatures and the courts to create new institutions for specifying and safeguarding the value of regulated property, which allowed contracts to supplant vertical integration in pipeline relationships. Following North's lead of systematic historical analysis, this volume looks to the source of such institutions as one of the roots of twenty-first-century pipeline industry structures.

Intangible Property Rights and New Pipeline Transport Markets

Pipelines are long, inanimate tubular steel assets accompanied by the occasional pump or compressor. There is nothing subtle about them. There *is* subtlety, though, in seeing pipelines not just as tangible steel tubes but rather as the physical means for providing an intangible property right to transport fuel from one point to another at a highly predictable payment to the pipeline owner. Like a tenant's intangible rights to commercial office space that accompany the payment of rent as enshrined in a long-term property lease (to occupy, divide, or sublet), pipeline "leaseholders" on the twenty-first-century gas pipeline system in the United States have similar rights. For the payment of the investment-cost-based regulated tariff to their supplying interstate pipelines, such pipeline users can schedule gas shipments for their own use. But they may also transfer their rights to transport gas between the points defined in their pipeline contract for any period they wish—from days to the full extent of their essentially perpetual contract with the supplying pipeline. Although these leaseholders pay regulated prices for their leases, they sell at unregulated prices on a transparent web-based trading site that lists all shippers, capacities, trades, and prices on a virtually instantaneous basis. It is a pipeline transport market unique in the world. The great institutional economist of the first half of the twentieth century, John R. Commons, of the University of Wisconsin, said that "in modern capitalism, the most important stabilized economic relations are those of private property."[21] Competitive gas pipeline transport arose in the United States once the value of the intangible contract right to point-to-point pipeline transport became so well defined and predictable that it rose to the level of private property. Familiar and exacting cost-based regulatory institutions for US interstate gas pipelines, developed in the mid-twentieth century, assure the underlying cost of that intangible private property. Other institutions formed late in the century to protect the value of that property and assure its frictionless, deregulated trade.

Those deregulated markets in intangible transport rights have enabled US regulators to tackle at once both problems that pipelines pose—to keep investment flowing despite the burden of asset specificity and to prevent harm to commodity markets. Gas pipelines in the United States no longer trade in the gas they transport or disrupt the vigor of spot or futures markets in gas. Gas producers and consumers of gas (and their agents) deal directly with one another in a universal and competitive spot market. They arrange for pipeline transport separately.[22] A robust futures market accompanies that spot market, just like the futures markets for other competitively traded commodities. Unregulated markets decide who uses the pipeline system on any day, and rivalry among many different potential pipeline capacity developers decides where and how the system expands to create more of those intangible rights without having to appeal to the judgment of the regulator. Against all preconceptions and after a century of struggle, the pipeline transport of gas across the continental United States has shown itself to be a structurally competitive business.

Such a role for US gas pipelines is an industrial manifestation of an insight that Coase dropped on his highly skeptical colleagues in 1960—which contributed to his being awarded the Nobel Prize just over three decades later. Coase saw the control of property rights (not simply the holding of assets) as the fulcrum of economic organization when he wrote that a "private enterprise system cannot function unless property rights are created in resources, and when this is done, someone wishing to use a resource has to pay the owner to obtain it."[23] With the property right having been defined, according to Coase, "chaos disappears" and so too does the necessity of government action to restrain those who would wish to use a limited resource (except for maintaining the legal system needed to define and enforce those property rights). Coase convinced his peers that it takes property rights to endow a resource with institutional scarcity in order to form the basis for trade and that a market could form where none had existed before simply by creating and safeguarding that scarcity value.[24] This was a radical notion in 1960, one that ran counter to long-accepted principles.[25] "Coasian markets" have since formed in pollution rights, carbon allowances, radio bandwidth, and other commodities through the creation and clear definition of intangible property rights comprising bundles of specific legal entitlements.[26]

A deregulated Coasian market for intangible inland gas transport rights exists and flourishes in the United States. The role of the federal regulator has changed to include safeguarding intangible capacity rights and the means for frictionless trade—and has accompanied a substantial reduction in traditional regulatory litigation and intervention over cost-based pipeline tariffs.[27] The extensive gas pipeline systems in Canada and Argentina have industry structures that would facilitate such a market, but those countries have not taken the step of defining the intangible rights or enforcing the markets that would permit Coasian markets for pipeline transport. The continental European and Australian gas pipeline systems are even further away from such a market in transport capacity rights, facing formidable institutional and political barriers to such markets.

The US oil pipeline transport system has nothing resembling a Coasian market in transport rights and no near-term prospects for achieving one. The root cause is its peculiar 1906 governing legislation dictating "common carriage"—which does not permit the rigorous definition of intangible rights upon which a Coasian market depends. Electricity transmission, for its part, is blocked from such markets not by such legislative fiat but by physics. Speed and unpredictability of electricity flows on interconnected power grids (among other things, moving at fifteen million times the normal speed of gas or oil in pipelines) defeat the creation of intangible point-to-point property rights over lines of known costs, under current technology.[28] Trucking and rail carriers have increasingly employed deregulated contracts to permit innovation and efficiency on long-standing rail and road networks. But rail and road common carriers deal with complex issues of terminals, interconnections, diverse products, reverse hauls, rolling stock, and competition from private carriers not subject to regulation.[29] And in any event, railroads and highways are networks with diverse users. The conditions of competitive opportunities for creating markets in intangible transport rights on these other modes do not appear.

Pipeline systems exhibit physical simplicity and predictability of operation, transport standardized products, and lack the unpredictable effects on other parties that would impede the creation of reliable rights for point-to-point transport. Pipeline systems are uniquely suited to the creation of the bundles of legal entitlements that form the basis of Coasian bargaining. Much of this volume examines the events—only sometimes intentional, often enough serendipitous—that allowed this sophisticated market to be grafted onto such an old and low-technology business with its baggage of aged governance institutions. But it works. It also raises the obvious question of whether competitive pipeline transport can spread to other regions in the world characterized by cartelization, entry-deterring behavior by incumbent gas suppliers, and stunted fuel markets.

Collective Action as a Lever for Regulatory Policy

The Coasian markets in gas transport that appeared in the United States at the conclusion of the twentieth century did not arise by themselves. Nor were they simply the result of keen regulatory action sprung from talented and inquisitive federal regulatory economists, judges, or commissions. Rather, they arose as the result of a decades-long legislative and regulatory conflict over the price of gas between gas distributors and northern US states on the one hand and gas producers (mainly the major US oil companies) on the other hand. Later these same distributors battled with pipeline owners for almost two decades more, on behalf of their millions of connected ratepayers, to transform traditional regulated pipeline services into competitive Coasian transport rights.

The interest among economists in how pressure groups bend legislation and public policy in their favor goes back to Adam Smith, who observed that "whenever the legislature attempts to regulate the differences between masters and their workmen, its counsellors are always the Masters."[30] But according to the late Mancur Olson, from the University of Maryland, Adam Smith's insights into collective action and its consequences were ignored until recent times.[31] Olson published an influential book in 1965—a veritable economic best seller—entitled *The Logic of Collective Action: Public Goods and the Theory of Groups*.[32] He overturned what had been accepted economic wisdom about how groups behave in the economy. Economists and political scientists had held that if everyone in a group (of individuals or firms) had some common interest, then the group members would tend to seek to further that interest. Simple enough but fundamentally wrong, as Olson showed. Using game theory and simple welfare-maximizing economics, he demonstrated that the larger a group becomes, the less incentive any individual member has to spend any time or money pursuing common objectives. In other words, small groups (of oil companies, say) can be effective in pressing their interests, but large groups (of millions of gas consumers, for example) will not.[33]

Legislation or regulatory policies that leave small groups of oil companies or pipeline owners (or a single trade group) facing uncoordinated masses of consumers gives the upper hand to the former in the drafting of regulations and the implementation of policy. This was the case for oil pipelines in the United States in the early twentieth century and remains true for Europe and its gas pipelines today. Without organized pressure groups of gas distributors, and their legal and economic advisers, the Coasian transport markets in the United States would not have evolved. Where those distributors came from, and how they prevailed, illustrates collective action at work.

Whether the users of a pipeline system can form effective pressure groups—small in number, well funded, influential—is critical to countering the public policy interests of pipeline owners. But the power of such buyer groups seems to have been overlooked by regulatory agencies and legislatures. The creation of the powerful and successful pressure group of regulated gas distributors in the United States was an unintended by-product of 1930s legislation aimed at breaking up abusive multistate utility holding companies. Policies intended to compel millions of small gas consumers to participate directly in gas markets (under the laudable-sounding label of "full retail access") defeat the prospect for such a balance of pressure groups in Europe and elsewhere. In all cases, the prospect for more competitive pipeline market structures is heavily dependent on whether and how such pressure groups can form to press their collective interests.

Plan for the Book

Throughout this book, pipelines serve to illustrate such an interdisciplinary approach to economic theory. The analysis will turn to the oil pipeline system in the United States or gas pipeline systems in various countries to illustrate the cause and effect of varied institutional environments. All of these pipeline systems are subject to: (1) neoclassical economic concerns (the economics of natural monopoly and government attempts to limit anticompetitive behavior through price regulation), (2) transaction cost economics (the seemingly inexorable pull of vertical integration owing to asset specificity), (3) institutional detail and historic path dependency, (4) the definition and safeguarding of intangible property rights leading to competitive markets, and (5) public choice and collective action in the shaping of pipeline regulatory legislation.

Tackling a century of worldwide pipeline development demands some organization. The book begins by focusing narrowly on the cost structure of the industry (which is more or less common to all systems) and the basic regulations that governments have imposed on investor-owned pipelines. The focus widens in the later chapters to analyze how transaction costs, property rights, evolving institutions, and collective action have shaped continuity and change in the industry. Whether the pipelines examined in this book ship oil or gas is not important—what is important are the institutions that govern their actions and industrial organization. The book's various chapters serve to highlight the reciprocal benefits of using the theoretical insights of the new institutional economics to analyze the pipeline industry while employing the industry to illustrate the power of the insights of the latter-day institutionalists.

Chapters 2 through 4 describe the traditional neoclassical economic analysis of pipelines and the regulations that developed around the world to deal with pipelines as presumed natural monopolies. Chapter 2 reviews some background elements for this study of pipelines: the existing literature on pipelines, the origins of the unique role of private capital in pipeline development in the United States versus the rest of the world, and a brief preview of where the analysis of pipeline markets in this book is headed. Chapter 3 examines the theory of natural monopoly, in both its static and multiperiod manifestations, to show how pipelines are highly idiosyncratic cases of natural monopoly, where the sustainability of their market power is challenged by geology, geography, time, and the regulation of pricing and entry. Chapter 4 describes how pipeline regulation responded to deal with the perception of market power held by investor-owned pipelines. These chapters together show that there are severe limitations on the ability of neoclassical economic analysis alone to provide traction for understanding the observable diversity in pipeline market development, regulation, entry, and pricing around the world.

Chapters 5 through 7 look at pipelines through the lens of the new institutional economics. Chapter 5 examines transaction cost economics and includes a review of the restrictions on transacting imposed by requirements for common carriage and third-party access (TPA)—terms too often loosely applied by economists. Chapters 6 and 7 show how the application of two different forms of transport regulation—first common carriage and later private carriage—led to vastly different kinds of industry structures for pipeline transport. One hundred years ago, US oil pipelines were saddled with common carriage regulations modeled on those used in the nineteenth century to regulate railroads, canals, and stagecoaches. That maladapted rein on the US oil pipeline industry has never been removed. It caused oil pipeline companies in the United States to vertically integrate and to devise various subtle methods to avoid the risks posed by asset specificity. Private carriage for US gas pipelines, on the other hand, provided the foundation for the eventual development of Coasian bargaining for legal transport entitlements.

Chapter 8 examines whether the efficient Coasian market in rights enjoyed by the US gas industry has any chance of developing in other pipeline markets—for either US oil pipelines, Canadian gas pipelines, or any of the other major gas pipeline systems around the world. It examines the stunted nature of gas commodity markets in Europe and Australia, which remain dominated by long-term, oil-indexed contracts, and it discusses how regulation would have to evolve in order for transport competition to appear, with the fuel supply security that goes with it. Chapter 9 returns to the broader economic, regulatory, and political issues illustrated by pipelines and reexamines what a century of pipeline problems reveals.

CHAPTER TWO

Existing Pipeline Studies and Private Pipeline Capital

Oil and gas pipelines are peculiar concentrations of capital. Often underground and unseen, they traverse the countryside, transcending natural and political barriers. Properly cared for, they last a very long time—fifty-year lifetimes are common. Pipelines have a vital mutual dependency on other businesses at each end that are themselves a heavy concentration of capital: producing wells, refineries, gas distribution utilities, industrial firms, and power plants. Pipelines represent quite unexciting technology that has changed very little in decades—big, dumb, tubelike inanimate objects. Just about everybody in the developed world buys products that depend on pipelines and reflect their cost, from natural gas to heat homes and businesses, to gasoline to fuel automobiles, to electricity from gas-fired power plants.

Yet, pipelines are nothing if not controversial. Russia has used its pipelines as a political lever against what it considers unruly former Soviet states. In the United States, President Richard Nixon had to sign a special authorization in 1974, in the wake of the OPEC oil embargo, for the Trans Alaska Oil Pipeline to overcome a phalanx of environmental objections (a proposed Alaska gas pipeline through Canada has been a political issue since the late 1970s). Widespread popular disgust at the monopolistic practices of pipelines led to congressional legislation both at the start and in the middle of the twentieth century that continues to govern US pipelines today. All around the world, pipelines are treated like monopolies that require regulation or government ownership. Perhaps no other industry combines such pedestrian technology and operation with such high-stakes politics.

There is no ambiguity about where an oil or gas pipeline runs or what it does: transporting fuel from one location along a defined path to another location. Indeed, it is a misnomer to call a nation's or continent's pipeline system a *network* as one uses the term for telecommunications networks or electricity transmission networks (where electrons flow at the speed of light in unpredictable ways). Pipelines are really no more networks than the collection of ropes on a square-rigged ship is a network. In both cases, every line has a definitive function. Despite this clarity of distinction, however, it is very hard to motivate private investor capital to build pipelines. Outside the United States and Canada, every major gas pipeline, and most oil pipelines, were built by governments. Even in the United States, the building of pipelines was a challenge for investors that depended on particular industry structures or purpose-built financial instruments created by insurance companies after World War II.

Pipelines are the quintessential choke points on trade—potential tollgates lying athwart trade routes. Oil and gas do not move over land very far without them. It was one Samuel van Syckel, who went to northwestern Pennsylvania in 1864 with the intent of dominating the shipping in oil around aptly named Oil Creek by cornering the market on barrels, who discovered teamsters, not barrels, to be the essential barrier to transporting oil. With that, he envisioned that one pipeline could displace thousands of teamsters hauling five to seven barrels of oil each in two-horse wagons over muddy western Pennsylvania roads—and that once those sweating, swearing teamsters and struggling teams left the business, he could treat as captives the independent oil producers in the region. Ever since van Syckel's realization, pipelines have made splendid monopolies, if not naturally then by the careful design of those who own them to shape public policy in their favor. As a result of this penchant for monopolization, every major pipeline that is not government owned is government regulated (sometimes both).

The world's pipelines have to respond to a great diversity in regulatory regimes—different services, different prices, and different rules for building new capacity. Some countries regulate their pipelines as common carriers. Others regulate pipelines as large-scale public utilities. Two pipeline systems, in the United Kingdom and Victoria, Australia, have for years been regulated as virtual storage tanks (where gas enters at one spot and almost magically reappears at another with no mention of the actual pipes in between). Gas pipeline companies in the United States are regulated as the builders and operators of a system of well-defined, long-term legal transport entitlements owned and traded at deregulated prices by others. In that US system, the ability of gas pipelines to act like monopolists—to exercise the inherent market power that van Syckel had discovered—has vanished. Each of these diverse approaches reflects the institutional history of pipeline development within the countries and regions whose pipeline infrastructure continues to shape their markets for fuel.

Existing Literature on Pipelines

It is evidently difficult for economists to study the world's pipelines, for nobody has yet done a systematic job of it. There are good economic studies of roads, railroads, and other forms of transport.[1] There are good economic studies of markets for electricity generation and telecommunications, including advances in the web.[2] There are also good economic studies of the global oil industry and the gas commodity industries in the United States, Europe, and elsewhere, some of which are cited in this book. But there are no existing economic studies pertaining to the world's *pipelines* as inland transporters distinct from the fuels they carry. Why not?

There are two likely reasons for this gap. First, the business of pipelines is deceptively simple. The technology involved in pipelines has not changed in many decades, leading to the perception that there is nothing new for economists to study. Second, pipelines have been in existence so long that the institutions that have grown up around them are complicated and rooted in controversies that took place up to a century ago.[3] The result is that we observe a highly diverse set of pipeline undertakings, both within the United States and around the world, despite the simple physical similarities of the pipelines themselves. The predominant neoclassical economic analysis of the latter half of the twentieth century does not provide a foundation to analyze the sources of such diversity—hence, the economic literature is mostly silent on the subject of pipeline transport.

The existing economic studies of pipelines have generally been a byproduct of the study of oil or gas commodity markets. For example, Paul

MacAvoy, of Yale, along with his academic adviser Morris Adelman, of MIT, and other collaborators (such as Stephen Breyer, now a US Supreme Court justice), have written extensively on what they considered the perils of regulating gas commodity prices and the inefficiency of federal gas industry regulations. MacAvoy wrote three books on the subject, but with all three the major focus was on gas, not pipelines (and not pipelines outside North America).[4] Similarly, Arlon Tussing, of the University of Alaska, wrote two editions of a well-regarded book on the US gas industry.[5] The books were historically informative and included tables related to the construction of pipelines in the United States. Nevertheless, as with MacAvoy, Tussing's main focus was the gas industry in the United States, not pipelines. Others have written extensively about the gas industry in Europe or elsewhere. In 1980, Malcolm Peebles wrote a highly informative history of the gas industry, from the Roman Empire to the era of worldwide shipments of liquefied natural gas (LNG). He covers both Europe and to a lesser extent the United States.[6] But as with MacAvoy and Tussing, his major focus is on gas rather than pipelines. Robert Mabro, of Oxford University, has long contributed to the literature on Europe's gas market, and his collection of papers, coedited with Ian Wybrew-Bond, represents excellent analyses of the sources of gas that now supply Europe.[7] As was the case with the book by Peebles (who contributed to Mabro's edited volume as well), this volume about Europe's gas industry deals with pipelines only tangentially, if at all.

To find scholarly studies that focus primarily on pipelines, one must look first to business historians. Two have written splendid histories, having been aided greatly by obtaining privileged access to internal pipeline company records. The first is Arthur Menzies Johnson, from Harvard Business School, who wrote two comprehensive studies of the birth and development of US oil pipelines from their genesis in the 1860s through the 1960s.[8] Johnson's analysis of the political events surrounding the birth of the world's first pipeline regulations—the Hepburn Amendment in 1906—is well researched, as is his analysis of the tension that developed between the US Department of Justice and the pipeline industry beginning in the 1930s.

The second is Christopher Castaneda, a business historian at California State University–Sacramento. In three highly informative books— again written with privileged access to corporate documents—he explored the development of the pipeline industry both before and after Congress passed the Natural Gas Act in 1938.[9] Castaneda captures the essence of the personalities, politics, and industrial disputes that accompanied the often rough-and-tumble years of gas pipeline development. Yet neither Johnson nor Castaneda is an economist, and they did not address root sources of market failure in the business, nor did they analyze the regulatory actions taken by governments against the palette of alternative methods of curbing pipeline market power while at the same time facilitating the flow of private capital to the business.

Legal scholars have also contributed to the literature on pipelines; two in particular are Eugene V. Debs Rostow, of Yale University, and George S. Wolbert Jr., of Washington and Lee University (later general counsel of Shell Oil Company). Both produced major studies around 1950. Rostow sought an alternative to common carriage for oil pipelines and the vertical integration it seemed to cause in the US oil industry.[10] Wolbert generally defended oil pipeline industry practices in reply.[11] Another legal scholar, Richard J. Pierce, from George Washington University, made a significant contribution to the literature on the evolution of the gas industry in the United States, with a focus on pipelines.[12] These are not economists either, however, and the debate between Rostow and Wolbert, for example, not only is dated but also takes on more of the appearance of opposing legal briefs than an economic analysis of pipelines.

The congressional regulation of pipelines has also attracted the attention of political scientists. M. Elizabeth Sanders, of the Department of Government at Cornell University, produced an insightful analysis of the Natural Gas Act of 1938.[13] She combined an economic view with a pluralist political perspective to analyze how Congress chose among competing constituencies to write the act and deal with the subsequent dispute-ridden decades. Sanders is sensitive to the role of pressure groups in bending pipeline legislation their way—as with any legislation in the United States. One cannot read her study without acknowledging that the public policies ultimately leading to the US gas pipeline industry's Coasian bargaining were not the work of economists but of politicians weighing the positions of competing interest groups. In the greatest fight of the era after the passage of the Natural Gas Act, each interest group had its own estimable economists, chief among them Morris Adelman, of MIT, for gas producers (the major oil companies) and Alfred E. Kahn, of Cornell, for gas distributors (representing millions of gas consumers). The laws Congress passed (or resisted, in the case of calls for deregulation of gas prices in the 1950s) were not the result of a search for economic efficiency as such. Rather, they represented the desire of leg-islators to accommodate the broadest interest groups that had prepared articulate and well-reasoned points of view.

Economists have studied pipelines too, but many of the most detailed analyses are almost fifty years old. In 1947, Emery Troxel, of Wayne University, wrote what is still one of the finest books on the nature of US regulation. His book and related published papers are important contributions to the economic analysis of gas pipelines and their regulation in the United States.[14] Leslie Cookenboo, of Rice University, wrote a dissertation on the economics of oil pipelines under Adelman at MIT, which he published as a book in 1955.[15] To modern eyes, Cookenboo seems unduly preoccupied with neoclassical concepts of natural monopoly—to the extent that he recommended giant mandatory joint ventures from oil-producing areas to maximize the amount of oil shipped and minimize its unit cost.[16] Kahn, in his famous 1971 *Economics of Regulation, Principles and Institutions*, devotes pages 152–71 of volume 2 to pipelines (twenty pages is significant for Kahn's comprehensive work)—having himself been involved in many of the disputes over the regulation of pipelines and gas prices in the 1950s and 1960s.[17] Over those twenty pages, in his soliloquy-like style, Kahn weighs the advantages of joint planning (which swayed Cookenboo) against the advantages of whatever genuinely competitive rivalry the pipeline industry could support—choosing not to side with one view or the other.

All of these economic analyses of pipelines, what there are of them, fundamentally came out of the conflicts between pipeline owners and the customers they served that were heard either before regulators or in the US courts. Outside of the United States and Canada, there *were* no major private pipelines until the end of the twentieth century. Subsequent to the privatization of many pipelines in Europe and elsewhere, the definition of private regulated property has remained somewhat vague, and disputes have been less likely to be heard in the courts. The newness of the privately owned pipelines, the less clearly defined limits on government involvement in their practices, and the uncertain split between EU and national jurisdictions for regulatory disputes in Europe have contributed to a shaky foundation for economic analysis there.

Ultimately, the existing body of economic analysis related specifically to pipelines, as a major inland transport industry, is thin. However, a common strain can be found among many of these works under the umbrella of the new institutional economics. While historical, legal, and some economic analyses of pipelines exist, there has been no analysis of the world's pipelines from that more varied theoretical perspective.

Pipelines and Private Capital

Private capital has played a key role in the development of pipelines. The privately funded origin of pipelines drives a wedge between the regulatory experience in the United States and almost everywhere else in the world. The most defining characteristic of US oil or gas pipelines is that they have all been financed by investor-owners under the assumption that each pipeline would pay for itself. Having a payment scheme in place from creditworthy parties for a new pipeline is a very big deal. It is the capital market's independent check on the wisdom of the line, its route, and size. Furthermore, such pipelines are private property, and the US Constitution prohibits any actions—by either the executive or legislative branches of government—that diminish the value of that private property without due process of law. Thus, no significant piece of legislation or regulation was ever introduced for US pipelines without its eventually being heard in the courts on the issue of the property interests of pipeline investor-owners.

The first large-scale transport project in the United States was the construction of the Erie Canal, to tie the "Northwest Territories" of the early nineteenth century (i.e., Illinois, Indiana, Michigan, and Ohio) to East Coast markets. Begun in 1817 and completed in 1825, the canal linked Lake Erie on the Great Lakes to the Hudson River, which flowed to New York City. As recounted by Lance E. Davis and Douglass C. North, because of the project's size and uncertainty, the infant nature of large-scale capital markets during that period, and the yet-nonexistent agricultural markets that the canal was designed to promote, government involvement in finance was a necessity.[18] The public seems to have accepted government participation in large-scale transport until the widespread commercial failures of such canal projects in the 1839–42 depression left taxpayers with the bill for projects that had far smaller expected net returns than the original backers had projected. The public's reaction was an initial refusal to meet state obligations, followed by legislation in the various affected states to prevent the recurrence of such political problems by legislative means—basically a prohibition on future public funding for such transport projects. Those reactions, which essentially raised the cost of public borrowing, were followed by the railroad-inspired capital markets of the mid-1850s forward—which lowered the cost and uncertainty of private borrowing. The canal experiences and the birth of very-large-scale private financing for the rail system generally ensured that other large-scale inland transport systems—oil pipelines and later gas pipelines—would not be government financed in the United States. The institutional evolution from public to private financing for inland transport in canals and railroads in the United States in the mid-nineteenth century was thus a reaction both to public opinion in unique circumstances and to a change in the relative cost of private versus public finance. By the time interest arose in constructing large-scale pipelines later in the century, it was thus preordained that the capital for such projects in the United States would come from investors and not the public.

Outside the United States, governments built major pipelines with public funds until late in the twentieth century. Drawing upon such funds, governments are able do what they wish with pipelines, as the financing constraint does not bind them (or discipline them) in the same way as it does investor-owners. Governments can build pipelines that investors would never support, and governments can charge shippers for pipeline services either more or less than those pipelines cost. Because pipelines last so long, the decision of governments on where to place pipelines and how large to make them has consequences for fuel markets and pipeline regulation long after pipeline businesses are privatized.

The dominance of private capital and the early role of independent joint-stock corporations in the United States versus other countries is not a story for this book. But in charting the source of US regulatory institutions, it is crucial to recognize the depth of support for investor ownership in utilities and other infrastructure businesses like interstate pipelines. In the early twentieth century, the role of private ownership of regulated businesses was studied and confirmed by a major and newsworthy national task force that sought to determine the wisdom of continuing the use of private capital for the building and operation of regulated utilities. Major players in the development of US regulation were part of this study, like economist John R. Commons, utility holding company pioneer Samuel Insull, and future Supreme Court justice Louis Brandeis. The study itself cemented the continuing role of private capital in regulated utilities and related US businesses.[19] After the study, the United States continued down the path of private utility investment while major pipelines in almost every other country were built and operated with public funds.[20]

The use of private capital for pipelines has facilitated the eventual transition to competitive point-to-point gas transport markets in two ways. First, the placement and size of pipelines in the United States had to satisfy the conservative perspective of financiers. Second, with private capital at stake, the Supreme Court would eventually decide how to value it for regulatory purposes.

The placement and sizes of the pipelines themselves generally responded to the fuels market to tie together producing and consuming regions without significant excess capacity. It is true that pipelines sometimes went places or made connections that seem to make little sense today. But an industry that required independent financing for every new pipeline after World War II had to satisfy insurance companies (which provided the bulk of the long-term loans that those individual pipelines would pay). Moving to a competitive transport market in entitlements would leave some pipeline capacity underutilized, as shippers subsequently revealed which transport links were valuable and which were not. But those problems, and the distress they caused pipeline companies, were relatively minor, as every pipeline project had had to satisfy the capital markets in the first place, meaning that few were heavily overbuilt.

The issue of the value of private property was equally important in the transition to a market in legal transport entitlements. In its epic 1944 *Hope Natural Gas* decision, the Supreme Court defined the value of private property that could not be taken from investors by the actions of regulators under the protections for property embodied in the US Constitution.[21] Without that bedrock base of property value in regulated gas pipelines, the kind of market envisaged by Coase would probably not be possible, as those legal transport entitlements would have an ambiguous cost base. As it is, they have a solid cost base that greatly facilitates their trade.

Starting with British Gas in 1986, many of the world's gas pipelines passed from public to private hands through various forms of privatization. Major pipelines in the rest of Europe, South America, and Australia were sold to private investors. But none of those privatized gas pipelines had to pass the private financing test that faced pipelines in the United States, and none of them can claim to embody the unambiguous property values that follow from a basic legal foundation like the *Hope* decision. The wedge that these two institutional issues drive between the advances in US gas pipeline regulation and the regulations in other regions, like the European Union or Australia, is substantial.

A Century of Wrestling with Private Pipeline Capital

A central goal of this book is to analyze how the dramatic transformation of the US gas pipeline system illustrates key economic theories of the new institutional economics—how a group of regulated interstate utilities selling delivered gas to captive customers transformed into an infrastructure system in which gas markets operate independently and freely and legal transport entitlements trade in virtually frictionless, web-based markets at deregulated prices. The Natural Gas Act of 1938 gave the federal regulators the duty to constrain pipeline market power and facilitate investment by regulating prices and market entry. The writers of the act would surely have been thunderstruck by the idea that, merely by altering the regulator's focus to include defining and preserving the value of long-term entitlements to point-to-point gas transport capacity, the source of pipeline market power would vanish and the seemingly endless trouble over how to choose among the firms petitioning for new licenses would generally take care of itself. The industry transformed over the course of about sixty years from traditional monopoly utility price and entry regulation to unregulated Coasian bargaining for acquiring transport rights in a highly efficient, and competitively expanded, inland transport market. The transformation happened as the result of two things, both of which involved the clash of interests over private pipeline property. The first was a long string of failures in trying other kinds of pipeline regulation. The second involved important advances in the legal, accounting, and administrative foundation for regulating the use and pricing of private property.

Congress's first regulatory experiment for pipelines involved the imposition of *common carriage* in 1906, with no defined accounting, price-setting, or licensing mandate by an Interstate Commerce Commission (ICC) preoccupied with railroad problems.[22] These oil industry regulations have been generally unsuccessful for more than a century. First and foremost, they failed to anticipate the primary lesson of transaction cost economics—the pull of vertical integration in the face of asset specificity. By failing to recognize that the mutual interdependence of pipelines and the customers they serve is inconsistent with a "common carriage" pipeline industry, that 1906 legislation inevitably pushed oil pipeline businesses to vertically integrate and form joint ventures. Other short-comings of that legislation are also brought to light through lessons now common in the new institutional economics. By failing to recognize that effective regulation of private property is impossible without a transpar-ent institution of mandated accounting, that legislation ensured that the source of oil pipeline prices would remain a mystery to outsiders for de-cades. By failing to give federal regulators licensing authority, that legis-lation left pipeline developers to compete with one another state by state for the eminent domain rights they thought they needed to push through new lines. In these respects and others, the 1906 legislation created eco-nomic obstacles that the industry and its regulators have necessarily worked around ever since—at times doing more to consolidate than re-strain the market power of the pipeline owners.

Having failed to deal effectively with oil pipeline market power, Con-gress tried again with gas pipelines in the 1930s, employing all that the country had learned about regulating investor-owned utilities since 1906. The country had learned quite a bit, as individual states like Wisconsin and New York developed new accounting and administrative institutions to deal with regulating their own private utilities and the Supreme Court addressed constitutional property rights questions that were central to the foundation for orderly government price regulation. The Supreme Court during this period freed Congress from any ambiguity regard-ing its responsibility to mandate accounting regulation, and it directed regulators on how precisely to respect property values when regulating tariffs.

The new legislation of the 1930s mandated austere separation of pipe-lines from gas distributors downstream, a strict accounting system, and rules for federal licensing if pipeline entrants could demonstrate "eco-nomic need." But these new, apparently highly advanced, regulations failed once again. Congress did not foresee how its utility-style regula-tion of interstate pipelines would disrupt the gas commodity market, as rival pipelines rushed to buy gas (to be paid for by captive gas distrib-utors) in a contest for new pipeline licenses. It was unthinkable in that context to permit pipelines to pay any amount they wished in a race with other developers to gain the necessary prerequisites for pipeline licenses. But it was also unreasonable to believe that the ponderous wheels of fed-eral regulation (which ground finely but slowly) could keep up with the market for gas—especially when much of the market lay within the states and outside of federal jurisdiction over interstate sales. The regulations of 1938 thus prompted producers to hold back gas supplies for in-state uses (thus outside of federal jurisdiction), Congress to temporize in only partially relaxing gas price controls, and pipelines to overreact by buying whatever supplies they could find at seemingly whatever terms were demanded by producers.

In the ensuing mess of the 1970s and 1980s, the regulator saw that a remedy—which it could not force upon pipelines under the 1938 law, reflecting a deep, underlying respect for the value of regulated private property under the US Constitution—was to have gas suppliers and shippers bargain directly without the pipeline as either a middleman or a common carrier. Trading a partial bailout for those distressed pipelines for "open access" to unaffiliated gas shipments, the regulator took the first step in the direction of Coasian bargaining by defining the transport entitlements that pipelines' traditional shippers would hold. But as long as pipelines shipped their own gas in competition with those entitlements, or could draw upon the embedded value of existing transport entitlements (in the form of a low historical cost asset base) to subsidize the building of new capacity, those entitlements would not trade effectively. It took more than a decade for the regulator to fully specify, through exacting and meticulous expert analysis, the definition of those gas transport entitlements that shippers could buy and sell on the frictionless exchanges that the pipeline companies were obligated to maintain.

It is unlikely that more than a small handful of the hundreds of analysts, lawyers, and regulators who worked on creating the market in well-defined legal gas transport entitlements had ever heard of Ronald Coase or knew of the requirements he had specified for such markets in the 1960 paper cited in his Nobel Prize.[23] The work involved in specifying the physical parameters of the transport entitlements, their embedded value in actual transport markets, and how they would trade was exacting and exhausting. But having completed that work in 2000, the gas pipeline regulator has transformed into a largely reactive agency. Gas pipeline rate cases are now generally perfunctory affairs, and licensing cases—once a source of tremendous Pigouvian controversy and regulatory effort—are uncontroversial.[24] While it had been a rough ride for pipelines, their customers, and the regulator, the process led to a highly competitive result in which pipeline companies build and operate the pipeline assets that provide genuine property rights in gas transport and shippers engage in

Coasian bargaining in their use and/or resale. Without a century of institutional work on how to square the constitutional rights of those holding private property in gas pipelines with the public's desire to be competitively served through those pipelines, the transition would not have happened. Such competitive pipeline transport never developed for the US oil pipelines—not because of any inherent industrial difference between oil and gas transport but because of institutional failures.

Before addressing the particular institutional elements that led to Coasian bargaining for gas transport capacity in the United States—and prevented it for oil pipelines—chapters 3 and 4 deal with regulation (or deregulation) as it has been shaped by a more neoclassical economic perspective. Chapter 5 continues with the essential contributions of the new institutional economics and the long evolution of these new markets.

CHAPTER THREE

The Economics of Production Cost: Pipelines as Natural Monopolies

Pipelines appear to be quintessential natural monopolies. Indeed, Archimedes of Syracuse, who mastered the concept of π in the third century BC, would have understood that the capacity within a pipeline was roughly π times a *squared function* of its radius. If the cost of the material for the pipe is roughly π times a *linear* function of that radius, then the average cost of pipeline capacity declines relentlessly with larger pipelines, as cost rises linearly while capacity rises exponentially. That means that a single pipeline is the least expensive way to serve the market for any conceivable quantity shipped—the definition of a *natural monopoly* for modern neoclassical economists. Archimedes may well have wondered why those future economists would make such a fuss over the concept.

But in the twentieth and early twenty-first centuries, the concept of natural monopoly caused a significant fuss as major oil and gas pipelines quickly became critical links in supplying the world's energy demands. Owing to natural monopoly, with larger pipelines having lower unit capacity costs than smaller ones, regulators and governments believed that major pipeline companies possess market power. With such market power, the theory goes, pipelines will not provide services efficiently—at competitive prices reflecting cost—if not government owned or price regulated. During the decade of worldwide gas pipeline privatizations in the 1980s and 1990s, the natural monopoly idea drove countries to privatize and regulate many of the world's gas pipelines as public utilities, as if pipeline competition and rivalry in inland fuel transport was out of the question. The perception of natural monopoly in pipelines has prompted some economists to favor building only one large pipeline from producing basins to major consuming areas as a compulsory joint venture.[1] Other economists perceive that the natural monopoly concept has been overemphasized to support pipeline regulation when an unregulated pipeline market would work better.[2]

Who is right—those who would treat pipelines as natural monopolies or those who would dispense with pipeline regulation altogether? It is a false dilemma flowing from a constrained point of view. The neoclassical economic perspective of production cost that leads to the concept of natural monopoly is insufficient to capture asset specificity—the key aspect of the pipeline industry that separates it from other businesses. But insufficient does not mean irrelevant: the cost structure of pipelines is an important part of understanding the nature of the business at particular locations and at particular points in time. This chapter looks at that cost structure and takes such traditional neoclassical economics as far as it will go to explain behavior in this industry—as much as anything to remove natural monopoly as an obstacle to a more useful institutional analysis. For despite the economics of natural monopoly, such a cost structure does not lead naturally to monopoly in real-world pipeline transport markets. Using practical examples from actual pipeline transport markets, it is easy enough to show that natural monopoly has failed in practice to produce a single, investor-owned pipeline in a major regional fuel market.

The Theory of Natural Monopoly for Pipelines

Economists are in wide agreement on the basic economic theory of natural monopoly. A statement regarding the theory appears in almost every elementary economics textbook. As in the case of perfect competition, however, natural monopoly is an abstract and static ideal, and there is no particular reason to believe that the textbook definition applies to real pipeline businesses any more than perfect competition applies in other markets.[3]

Are pipelines natural monopolies, either individually or as parts of interconnected systems? It is inescapably true that between two points, for a specified increment of demand, a single pipeline is less costly than two.[4] Figure 1 illustrates a traditional natural monopoly, where average cost declines throughout the entire range of output. If unregulated, a firm setting a single price for its services would produce the quantity at which its marginal cost equals its marginal revenue, or Y_m at P_m. Ideally, the regulatory agency, if there is one, would like to compel the firm to produce the competitive output, Y_c. But at that output, demand does not permit charging a price that covers average cost, and the firm could not survive without tax-financed (or other) subsidies. In this case, the best that a regulator can do to mimic a competitive outcome is to set a price equal to average cost, allowing the company to serve the demand at that price level, Y_r and P_r.

In the 1980s, William Sharkey, of Bell Laboratories, and William Baumol, of Princeton (and his coauthors, John Panzar, of Bell Laboratories, and Robert Willig, also of Princeton), closely examined the definition of natural monopoly and its sustainability—that is, a firm's immunity from the profitable entry of competitors.[5] Not only did Sharkey and Baumol generalize the definition of natural monopoly, but they also showed how firms may *not* be immune to profit-seeking entry, even with a business defined by a declining-cost technology such as that pictured in figure 1.

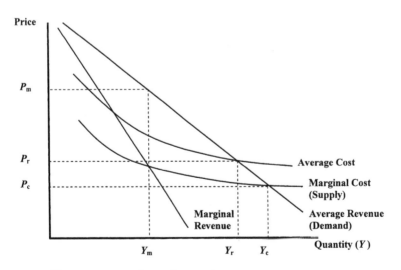

FIGURE 1. The cost structure of natural monopoly

In an effort to generalize the theory, Baumol and his colleagues described three functional definitions of natural monopoly. The first and most restrictive applies to figure 1. It specifies *declining marginal cost* (MC)—that is, that the marginal cost of larger quantities is always smaller than that of smaller quantities, precisely consistent with the MC curve in figure 1. If MC declines in this manner throughout the whole range of output, then average cost (AC) will decline throughout the range of output as well. There is a less restrictive case in which MC rises but is still below AC, causing the latter to still decline with output. In this case, the natural monopoly of the entity stems from *declining average cost*, where AC declines over the relevant range of output, whatever might be happening to the MC curve beneath it (as if the MC curve wobbled upward in the accompanying figure but nevertheless stayed below the AC curve).

There is a still weaker definition. Natural monopolies do not require AC to decline everywhere. In this weakest form, an industry is said to be a natural monopoly if, over the entire relevant range of outputs, the firm's cost function is *subadditive*.[6] This means that no matter what is happening to the cost curve, even if it exhibits increasing costs in short bursts, two or more firms cannot produce any relevant output more cheaply than a single firm. This is the weakest requirement for a natural monopoly in the production of a single product, and it embraces as special cases both the concepts of declining average and declining marginal cost.[7]

There is no controversy regarding these neoclassical economic definitions of natural monopoly. And to the extent that any of the definitions hold for a single-product firm over the relevant range of output, the natural monopoly is sustainable and immune to the threat of competitive entry. Whether any of these definitions apply to a single-product industry depends on whether the AC curve is ultimately U-shaped within relative levels of output (whether costs ultimately rise, signaling an exhaustion of economies of scale). A single-product firm with a strictly subadditive cost function is immune to competitive entry. But a firm with a U-shaped AC function, which none of the previous three definitions allow up to the relevant output level, may not be.

While it might be reasonable to believe that the average cost of pipeline capacity would decline throughout the relevant range of outputs, general perceptions can be wrong, and often have been wrong in the past when applied to regulated industries. For example, for much of the twen-

tieth century, the rule of thumb was that power generation stations exhibited costs like those in figure 1—that is to say, bigger power plants displayed ever-decreasing AC. The cost functions used in the econometric
analyses of the 1950s and 1960s allowed no U shape to the AC curve as
a point of simple mathematics.[8] At the time, there was no way to test a
natural monopoly hypothesis for the electricity generation industry. It
took the development of new econometric techniques, enabled by the
faster computers of the 1970s, to test empirically for natural monopoly.
In the first significant study utilizing the new theory and the more powerful econometric estimation methods available, Laurits Christensen
and William Green, at the University of Wisconsin, rejected the hypothesis of declining AC for power plants in the United States. They found
that by 1970 nearly half the electric output of power plants was produced
by firms under almost constant cost conditions. They also found a small
number of firms operating in the upward-sloping segment of the AC
curve, becoming the first economists to empirically estimate a U-shaped
cost function for any industry.[9] By 1970, the power plant rule of thumb
had been disproved.

Examining costs for pipelines is not as challenging as examining
costs for electric power plants. Power plants are complex electricity-
manufacturing facilities; pipelines are merely long steel transport routes
enhanced by pumps or compressors. In addition, the *operating leverage*
of pipelines (the proportion of fixed to total costs) is substantially larger
than for power plants, where fuel and labor represent more than half the
total cost. For pipelines, charting the cost curve can be reasonably ap-
proximated by examining the relationship between output quantities, or
the capacity of the line, and the construction cost.

The Observed Cost Structure of Pipelines

The cost of constructing pipelines has been a focus of industry and economic studies of pipelines from the start. Arthur Johnson examined extensive source documents from the Prairie Oil and Gas Company to
document the cost of building oil pipelines of various sizes. Prairie was
Standard Oil's purchasing, producing, and transporting affiliate in and
from oil fields in Kansas and Oklahoma. As those oil fields were far from
major markets in Missouri and Illinois, the company developed into one
of the major oil pipeline firms. In 1915, Prairie owned over five thousand

miles of trunk and gathering pipelines and was the largest pipeline system in the country. As of 1919, in the midst of an aggressive expansion program, the company's aggregate cost of pipelines of various sizes from two to twelve inches in diameter showed a relatively flat relationship in the cost per mile per inch of diameter. That is, while the cost per mile for Prairie's pipelines, from two to twelve inches, increased proportionally to diameter (from \$1,613 to \$19,533, respectively), the cost per mile *per inch of diameter* only doubled (from \$807 to \$1,628).[10] Since capacity increased as a squared function of diameter, it is clear that Prairie's costs declined consistently with figure 1.

The first systematic study of the cost structure of gas pipelines documented a similar declining relationship between average cost and capacity. Emery Troxel used Federal Trade Commission data from the mid-1930s to perform his own economic cost calculations regarding gas pipeline construction.[11] During this period, there was no federal regulation of interstate pipeline rates, and the Uniform System of Accounts that forms the basis for all economic regulation in the United States had not yet been applied to pipelines.[12] At that time, neither federal agencies nor states had gathered data on pipeline construction. Further, Troxel noted that there was no generally accepted line of demarcation between gas transmission and consumer distribution lines and field gathering lines—in a nation that by 1933 already had 65,543 miles of trunk natural gas pipelines.

Faced with difficulties in obtaining reliable gas pipeline data and in defining both gas transmission lines and basic pipeline costs, Troxel constructed his own data. Using his own assumptions and calculations regarding the contributions of capital and operating costs for pipelines, he found a roughly constant relationship between the diameter of gas pipelines and the cost per mile per inch of diameter, with a mile of 3.5-inch pipelines costing an average of \$1,320 per inch of diameter, much the same as the \$1,216 cost per inch of diameter of an 18-inch line.[13] The cost per inch per mile shows some diversity among pipeline sizes, reflecting the idiosyncrasies of building particular lines in particular locations. But, again, it was a roughly constant cost per inch of diameter consistent with figure 1.

This type of relationship has been confirmed by more modern sources. Trends in gas pipeline construction costs for 1954, 1959, and 1960, collected in pipeline filings before the Federal Power Commission (FPC), showed the same roughly constant cost in cost per mile per inch over a wide range of pipeline sizes.[14] The same is true for data based on Federal Energy Regulatory Commission (FERC) gas pipeline permit filings from 1980 and 1994.[15]

When it comes to transporting oil and gas over long distances, installed pipeline capacity is only one measure of the total cost of delivery. The longer the pipeline, the greater the need to increase pressure along the line in order to deliver acceptable pressures at the outlet. Capacity in this respect is achieved in three ways: greater pipeline diameter, insertion of pumps or compressors along the line, and "looping" (installing additional sections of pipeline alongside the original line) at strategic places to increase the effective diameter of the line. Pipeline engineers balance compression, basic pipeline diameters, and loops in an effort to deliver capacity, at acceptable outlet pressures, for minimal cost.

Sustainability of Cost-Driven Pipeline Natural Monopolies

What does economic theory have to say about the natural monopoly when it is generalized from the highly stylized, static model exhibited in figure 1? Pipelines expand incrementally over time to serve growing markets. In most markets, pipelines also exist in a diverse spatial landscape, where there are multiple ways of supplying particular markets from existing gas fields or import points. Economists have examined how natural monopoly is affected by both considerations.

Natural Monopoly in the Case of Pipeline Capacity Expansions

The multiperiod theoretical investigations of natural monopolies deal with whether monopolies are *sustainable* in production environments in which firms expand over time. In other words, a firm is an *intertemporal natural monopoly* if its product cannot be produced at a lower cost by any combination of two or more firms as demand grows over time.[16] Using a simple two-period model, Baumol and his colleagues were able to investigate several characteristics of sustainability of natural monopolies. To the surprise of many, they found that the incumbent natural monopoly could be *unsustainable* when it came to the point of expanding its system to serve growing demand. Worse, for the incumbent, Baumol showed that successful entry may lead to the actual displacement of the incumbent before it is able to recover its initial investment costs. The crux of the issue is the lumpiness of capital additions needed to serve growing demand. With an industry characterized by declining average costs like pipelines, adding capacity in bits and pieces over time will inevitably cost more than building the capacity required by the second period's entire demand all at once. In Baumol's two-period model, adding an increment of capital could disadvantage the firm relative to the cost of an entrant, unless the level of fixed cost needed for a new firm to enter the business is sufficiently great to cover the difference. In this respect, the entrant in this two-period model holds an option that the incumbent lost through its original entry—the option to build the whole, larger, system all at once. Put another way, waiting conveys a valuable option to a potential entrant that might, under certain conditions, kill an incumbent.

This model and others like it, however, are bound by highly restrictive assumptions. The firm in question sets a single price, has no information advantage over rivals, and does not respond to the threats of other firms by undercutting prices to prevent entry.[17] Further, the industry does not face regulation of pricing or entry. Also, the firm cannot form contracts with its customers or otherwise require multiperiod commitments that effectively remove a large part of the market from the possible entrant.

In the real world, the incumbent has many options available both to preserve its monopoly, at least long enough to pay for the capital costs of the business, and to help it gain an advantage over possible entrants in expanding capacity. Two of these alternatives are particularly important for pipelines: contracting with existing customers to remove portions of the market from available entrants, and entry regulation on the part of a regulatory body that may itself constitute an administrative barrier to entry.

To the extent that the incumbent firm forms contracts with existing customers, the entrant seeks only to serve the increment of demand rather than the entire market, thus raising its average cost of constructing an incremental pipeline. Price regulation may also help to assure the incumbent's position in the market. If a regulator sets prices to equal average cost during the period in question, including that period's depreciation of the invested capital, then when it comes to a subsequent second period, the incumbent has fully functioning but partially depreciated capital and hence lower regulated prices. In this event, the entrant must overcome the second-period pricing advantage represented by the depreciation paid by customers in the first period's regulated prices. In effect, the first period's regulated charges paid down part of the incumbent's regulated capital account. To the extent that the expansion of capacity comes after a number of years, the incumbent holds a powerful regulated price advantage over the potential entrant. This effect will be more pronounced still—a veritable entry blockade—if the regulator uses nominal (i.e., noninflated) historical accounts for the purpose of setting regulated charges. In both cases, the practice of setting one average charge for all capacity—original and expanded—generally destroys the competitiveness of the potential entrant in subsequent periods.

The theory thus provides no particular reason to question the sustainability of natural monopoly for a particular pipeline in a multiperiod model. The restrictions are simply too strict and unrealistic for the natural monopoly of a particular pipeline from one point to another to be unsustainable, allowing the entrant the greater flexibility to serve the whole market economically. That is, the value of the option to wait for the market to develop, which drives the competitiveness of the entrant in other models, is not sufficiently strong to overcome the cost-based advantage of incumbents in a market with cost-based regulated regimes.

Pipelines as "Networks"

This book avoids using the term *network* to describe pipeline transport systems. For telecommunication, electricity transmission, road/rail transport, or even locally distributed gas systems, *network* conveys a particular idea that does not apply to long-distance oil or gas pipelines. For telecommunications and electricity transmission, *network* refers to a connection between users that happens at the speed of light and is either untraceable or unpredictable. Indeed, the inability to predictably trace electrons on electricity transmission systems prevents the creation of physical contract paths for electricity sales. The lack of such contract paths, which would otherwise tie electricity sales transactions to particular transmission lines, has driven the competitive reform of electricity markets to require that power be sold into "pools" that are managed by pooled transmission systems, which are themselves overseen by "independent system operators." This is not so for oil and gas pipelines, in which the flow of the fuel is predictable from point to point and easily accommodates physical contract paths, tying particular fuel sales to particular pipelines.

Railroad, highway, or airline transport systems deal with simultaneous back-and-forth demand in addition to handling a diversity of freight and passenger traffic.[18] The planning, construction, and operation of these transportation systems do not revolve around the one-way shipment of commodities from wells to consumption centers that typifies oil and gas pipelines. Local gas distribution companies maintain complex systems of pipes going up and down city streets in metropolitan areas, and it might be possible to call such a system a network, particularly in that such distribution companies create redundant links in their systems to provide security to system users in case of a break.

In contrast, oil and gas pipelines are built from point to point to handle a predictable flow of fuel. Even on the extensive, century-old oil and gas pipeline systems in the Appalachian region, there is no ambiguity about the origin and destination of the oil and gas. The economic and practical complexities of transacting associated with those other "networks" (such as network externalities, reverse haul, and diversity of freight and passenger traffic) do not apply here.

To be sure, there are complexities unique to oil and gas transport systems. Oil products pipelines, for example, handle different types of petroleum products (heating oil, jet fuel, gasoline, etc.) as "batches" on a regular revolving schedule. One batch pushes the other through the pipeline, and tanks at either end of the line separate the different products. Most gas pipeline systems, like that in the United States, require certain upper and lower calorific limits for gas entering their lines, but some gas systems (such as those in the Netherlands or Poland) handle more than one type of gas. There are complexities associated with managing product pipelines or dual-gas systems, but they are not like the complexities associated with telecommunications, railroad, highway, or electricity transmission systems. The lack of network economies makes any lesson from network industries unlikely to apply to the pipeline transport industry. Avoiding calling gas and oil pipelines networks simply helps to avoid inapt comparisons.

The Practical Empirical Weakness of Pipeline Natural Monopoly

Pipelines obviously have a sharply declining cost structure in the planning and expansion stages, which the basic neoclassical economic theory is equipped to handle. Where that economic theory struggles, however, is in whether and how major gas pipelines sustain their naturally monopolistic position in the world—particularly on the scale of the continental gas systems in North and South America, Europe, and Australia. While between any two points there are manifest economies of scale in employing the largest diameter pipe available, the sources and markets—as they grow and shift—provide ample possibilities for competition between existing and prospective competing lines.

Three attributes of pipelines get in the way of a clean transition from the theory of natural monopoly, driven by economies of scale, to the pipelines seen in practice. The first is geography and the locationally fixed and irretrievably sunk nature of pipe costs. Natural gas comes from geological formations that steadily decline. They are what the trade calls a "wasting asset." As recoverable gas fields are depleted, gas production regions move around and evolve, while gas pipelines, reflecting non-redeployable capital, do not. Most of the original gas fields of Appalachia, which fueled the growth of the gas industry, are now good only for gas storage. When new fields develop, or when new markets arise, there is no guarantee that the right-of-way of an incumbent pipeline will provide any particular cost advantage over a rival new entrant.

The second attribute is politics, and the way it determines who builds investor-owned pipelines and where they are built. The competition for new pipeline licenses examines considerations beyond cost alone. For example, land is the ultimate scarce commodity, and major gas pipelines need lots of it, often requiring government approvals for rights-of-way and permissions to build on or under private land. There has always been a contest between cost, convenience, the lure of new gas pipeline promoters, and the desire to foster some type of long-term rivalry among gas pipelines. The natural monopoly view that a single regulated gas pipeline company could serve at lower cost has often failed to sway policy makers and regulators.

The third attribute is regulated pricing itself. The straightforward theories of natural monopoly examined earlier in this chapter assume uniform pricing, which is not a universal practice. Indeed, federal regulators in the United States discovered that requiring the segregation of costs and the separation of prices for different vintages of capacity was necessary for the creation of transport markets. Such "incremental pricing" would prevent the incumbent pipelines from engaging in an insidious form of price discrimination by drawing upon the market value of old capacity (by charging existing users more—called "roll-in") to price incremental capacity below its cost, undercut rivals' prices, and hence effectively bar entry.

Natural Monopoly for Pipelines Challenged by Geography and Geology

The pipeline industry is a slave to geography. The oil and gas fields are where they are, and major import or processing points are comparatively uncommon. It may seem elementary to say that the pipeline industry links these supply points to the consuming markets, but herein lies one of the major features undercutting the industry's ability to be characterized as a natural monopoly. The major consumption markets often lie between different supply areas, leading to intensive rivalries between producers and pipeline companies striving to meet the demand. Furthermore, the first major pipeline to supply a metropolitan consuming market is often left virtually stranded by the discovery of nearer and larger supplies. The early history of unregulated gas pipeline development in the United States has many examples of small gas fields (and the pipelines to those fields) bypassed or rendered valueless by the development of larger and more economical gas supplies flowing from newer fields.[19] There is no better modern illustration of the threat that geography and production geology play in disrupting pipeline market monopolies than when a new gas discovery trumps an existing gas pipeline development to an older, more distant field. The Argentine and Australian markets provide two examples of this phenomenon.

ARGENTINA: BYPASS OF THE TIERRA DEL FUEGO PIPELINE. Argentina has one of the world's oldest and most extensive gas pipeline systems. The country's first major pipeline, a 1,900-kilometer, ten-inch pipeline built from Patagonia to Buenos Aires in 1949, was by far the longest non-US pipeline at the time.[20] In the 1960s, Argentina discovered substantial natural gas reserves in Tierra del Fuego's Austral basin. Tenneco, the US company operating the Tennessee Gas Pipeline Company, won a contract to build a 3,568-kilometer pipeline from Tierra del Fuego across the Strait of Magellan and up the coast to Buenos Aires. The pipeline was a major engineering achievement, crossing many rivers and bays. It commenced service to Buenos Aires in 1965.

In 1991, the government of Argentina moved to privatize Gas del Estado, the vertically integrated state gas transportation and distribution monopoly. By that time, there were three sources of gas serving the major gas markets around Buenos Aires. In addition to the southern supplies from the Austral basin, a pipeline had been built to bring gas from Bolivia in the north. In addition, two major 2,000-kilometer pipelines were built in 1970 and 1988 to the more recently discovered Neuquén basin in the west. At the time of privatization, the Argentine government made a policy decision to divide Gas del Estado into several regional gas distribution and at least two rival gas pipeline transmission companies.

Two problems developed with the southern pipeline as a result of the multiple new, privatized pipeline companies. First, the incremental cost of maintaining the longer, older, more technically challenging pipeline to the Austral basin in Tierra del Fuego was very high. Second, the three major wells in the Austral basin fields had greater combined deliverability capacity than the southern pipeline had capacity to ship gas to the major market areas. The construction of reasonable, cost-based regulatory accounts for all the pipelines in Argentina raised the tariffs for the southern pipeline so high that they wiped out the "netback" value of gas in the Austral basin. Given a reasonable forward-looking price for transporting gas from the major basins, the Austral basin, with its aged 2,630-kilometer pipeline, could not compete with the newer 1,194-kilometer pipeline from the larger Neuquén basin on an equal footing. In terms of economic efficiency, the best plan for serving the major Argentine markets would have been to allow the southern pipeline to reach the end of its useful life and, rather than replace it, to develop the closer Neuquén fields and associated pipelines to fill the gap.

The "equal footing" pricing of the southern field, which could either wipe out the value of Austral basin gas or presage its eventual retirement as a producing basin for Buenos Aires, was considered highly impolitic by the government of Argentina (and would have outraged the southern provinces). As part of the privatization, the government chose to preserve the value of the southern gas fields by subsidizing the southern pipeline at the expense of the western pipeline to the Neuquén fields.[21] It was a political solution to a thorny economic problem caused by the newer, more economical western pipelines bypassing the original southern pipelines.

AUSTRALIA: BYPASS OF THE COOPER BASIN PIPELINE. For many years, the city of Sydney, Australia, was served by the coal-gas-based Australian Gas Light Company (AGL), an investor-owned utility. After the Cooper basin gas fields, in Australia's interior, were discovered and developed in the late 1960s, the Commonwealth government built a 1,299-kilometer pipeline to move that gas to Sydney in 1976—the "Moomba-to-Sydney" pipeline. AGL then converted to natural gas.

After the Cooper basin reserves were discovered and developed, a joint venture comprising Esso and BHP (an Australian mining company) discovered and developed gas fields in the Bass Strait, off the coast of Victoria, to the south. The gas was shipped to shore for treatment before being transported to Melbourne, where it displaced the manufactured gas used by the government-owned Gas and Fuel Company of Victoria. By 1995, there was heavy interest in bringing Bass Strait gas to Sydney. Not only was the Victorian onshore processing plant only 795 kilometers from the city, but the proven reserves in the Bass Strait were roughly three times the reserves in the Cooper basin.[22]

From a basic competition perspective, it was clear long before 1995 that, in the long term, Bass Strait gas could easily displace Cooper basin gas for service to Sydney. In 1994, however, the government enacted the Moomba-Sydney Pipeline System Sale Act, which sold the interest in the Moomba-Sydney pipeline to AGL, the local distributor in Sydney. In a practical sense, giving the distributor (the principal gas buyer) a major financial stake in one particular pipeline inhibits the development of competing lines and gas sources. The Bass Strait line, now called the Eastern Gas Pipeline, was eventually constructed by a consortium led by Duke Energy, despite AGL's desire to protect its economic interests by favoring the Moomba-Sydney pipeline over the new entrant. The new line, however, had to overcome high entry barriers placed in its way by AGL.

In any event, as occurred in Argentina, the discovery of larger gas basins close by inevitably impaired the value of the incumbent major gas pipeline and demonstrated that, despite their cost structure, major gas pipelines face persistent threats of rivalry.

Natural Monopoly for Pipelines Challenged by
Politics and Entry Regulation

While the early oil pipelines in nineteenth-century Appalachia were built with little or no intervention or approval from any governmental agency, modern pipelines require regulatory licensing approval—for various reasons, the most important of which being the orderly acquisition of the right-of-way. Some regulatory bodies search for the lowest cost and the single largest pipeline. Other regulators have kept the door open to future firm rivalries by turning a deaf ear when incumbents insist that they can expand at a lower cost.

The subject of competition for licenses (known in the United States as certificates of public convenience and necessity) to build new regulated pipeline capacity is multifaceted.[23] It pits those looking for scale economies against those seeking to advance new or innovative projects by competitive entrants, possible scale economies notwithstanding. The tension between these points of view is most apparent in Alfred Kahn's extended discussion of the subject from 1971, in which he examined both the centralized planning and the competitive point of view.[24] Various administrative law judges and commissions found themselves deciding whether to allow the licensing of new gas pipeline projects despite complaints by existing gas pipeline companies that the new projects were duplicative or could be viewed as inefficient.[25]

What is striking about these cases is the seeming judicial preference for competitive entry over vague or long-term arguments from incumbents or joint ventures based on economies of scale.[26] Judges had a tendency to side with the entrants over elaborate efforts on the part of incumbents to make those entrants prove that they were the cheaper option or that the high cost of the incumbent's operation was not their own fault.[27] It is difficult for regulatory institutions to deal effectively with all of the relevant issues involved in licensing, particularly in the face of persistent and creative incumbent pipeline interests. As a result, judges were willing to defer to a competitive rival in hand rather than follow the merely conceptual path of the greatest possible scale economies.[28]

The Ephemeral Natural Monopoly in Pipelines

Decades of empirical analyses show that, in the static planning sense, pipeline costs for individual lines decline relentlessly in the face of larger pipeline installations. But this aspect of declining average costs—the archetypal neoclassical natural monopoly—is often enough not a practical description of actual pipeline markets. Long-distance pipelines do not have the characteristics of gas distribution companies' local area networks, which effectively and "naturally" bar the entry of competitors. Major transportation pipelines face important barriers to the successful application of the cost advantages that would otherwise typify a natural monopoly.[29]

Ultimately, the concept of natural monopoly cannot advance the economic analysis of pipeline transport very far. And even if there were some element of natural monopoly in a static sense, it would beg the question of whether rivalry among pipeline competitors may provide more efficient performance over time. Despite their static cost structure, pipelines are often highly ineffective natural monopolies. In a practical sense, they are frequently incapable of dominating large-scale pipeline transport markets, though many try to do so. In actual pipeline transport markets, the scale economies evident in pipeline cost structures dominate only if governments, or their regulators, allow it to happen.[30]

Regulating Pipelines as a Response to Monopoly

Monopolization, politics, and pipeline economics came together in the early twentieth century as President Theodore Roosevelt and his allies in Congress sought, in the era before effective antitrust enforcement, to break the monopolistic abuses of the Standard Oil Company. This marked the first time that pipelines came to the attention of Congress, as the Senate debated whether the power of the Interstate Commerce Commission (ICC) to regulate common carriers like railroads should apply to the pipeline industry. Ultimately, while Congress extended the ICC's jurisdiction to oil pipelines in the 1906 Hepburn Amendment to the Interstate Commerce Act of 1887, it exempted gas pipelines from those regulations, a critical move that made the twenty-first century's competitive gas transport system—with its Coasian bargaining—possible. But the twenty-first century was a long way off when the Standard Oil Company was using its unregulated oil pipelines to dominate US oil production.

In addition to the passage of the Hepburn Amendment, 1906 was a pivotal year for the development of new regulatory institutions in the United States. It was the year that two states, Wisconsin and New York, first passed legislation to regulate utilities at the state level; the other states did likewise soon after, following those two similar examples.[1] It was also the year of a major and much-anticipated study assessing the relevant attributes of privately owned versus publicly controlled utilities.[2] That study, published in 1907, was responsible for the continued reliance on investors, rather than governments, to build and own US public utilities. The year 1906 also marked the first time a university—the University of Wisconsin—offered a class on the economics of public utilities.[3] All of these efforts would portend the development over the next three decades of new legal, accounting, and administrative institutions for the economic governance of regulated public service companies.

With all of these regulatory initiatives happening at once, it is easy to see that Congress had little experience to draw upon to craft a workable Hepburn Amendment. The institutions of effective public utility regulation in the United States, at either the state or federal level, were still to be developed. There was no generally acceptable accounting system for regulatory purposes (that would come in the 1930s), no legislatively mandated administrative practices (they arrived in the 1940s), and no unambiguous guide for applying the US Constitution's protections for the private property used in providing public services (that would begin in 1928 and conclude in 1944). Anyone pursuing regulatory remedies regarding oil pipelines had only the railroad regulatory practices to guide them. The Interstate Commerce Act, patterned after the British Canals and Railways Act of 1854, attempted to codify and enforce a number of legal principles embedded in common law.[4] But through the turn of the twentieth century, the courts continued to hamper the ICC's attempts to stem abusive railroad practices, prompting the Roosevelt administration to encourage Congress to give the agency additional powers.[5] This chapter shows that institutional details matter—and history matters—in the analysis of markets for pipeline transport.

Standard Oil Company and the First Pipeline Regulation of 1906

The world's first petroleum pipelines displaced three thousand teamsters in the Pithole Creek area of northwestern Pennsylvania in the early 1860s. Those pipelines were vastly more efficient than existing methods of shipping oil, in which teamsters hauled wooden barrels of oil that were then transported via railroad tank cars. By attaining rights-of-way and constructing lines on particular routes or to particular locations, pipeline owners forestalled competition from these other modes

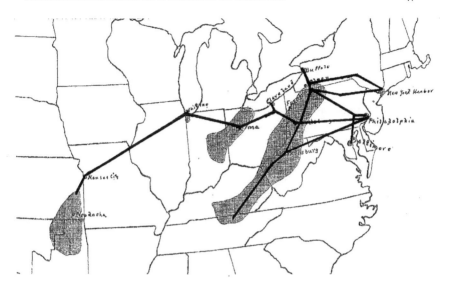

FIGURE 2. US oil pipelines in 1904, before the Hepburn Amendment (Reprinted from *Report of the Commissioner of Corporations on the Transportation of Petroleum* [Garfield report; Washington, DC: US Government Printing Office, May 2, 1906], facing p. 45)

of transport. In other words, the pipelines had the upper hand and could dictate terms to any oil producer in the vicinity.

John D. Rockefeller began trading oil for refining purposes in Cleveland in 1863. With a group of associates, he steadily built up a refining business until his Standard Oil Company (and its affiliates) owned or leased approximately 90 percent of domestic refining capacity by the late 1870s. By 1874, Rockefeller had observed and grasped the full potential of pipelines to bypass teamsters and compete directly with railroads in transporting oil from the fields to the major shipping points and markets. The extent of the crude oil pipelines controlled by Standard Oil, as of 1904, is shown in figure 2. It is useful to note in that figure how Rockefeller consolidated his early lock on oil transport over a relatively limited geography where the source of oil and its principal destinations were relatively close to each other.

Attempts at Oil Pipeline Regulation by the States

Early crude oil pipelines stayed out of the public eye and off the legislative agenda in Congress, which viewed pipeline problems as inconsequential outside the oil-producing states. Because it was not possible for independent producers to gain access to the regional pipelines, and there was no reliable way to gauge the amount of oil that those pipelines transported or stored, the oil producers in the region complained that their prices per barrel at the wellheads were being manipulated by Standard Oil's pipelines. Their allegations led to the world's first pipeline regulations—initial, halting efforts by the Ohio and Pennsylvania legislatures in 1872 to require eminent domain pipelines to act as common carriers. New York passed the first "free oil pipeline" law in 1878. Pennsylvania followed in 1883, and by 1906, more than twenty-one other states had placed pipeline, eminent domain, and explicit common carriage laws on their books.[6]

The state provisions, however, had little impact on the use of Standard Oil–affiliated pipelines by unaffiliated oil shippers. In a letter from the Standard Oil Company to Senator Henry Cabot Lodge of Massachusetts, a letter that Lodge read to the Senate in order to demonstrate "the futility of this [Hepburn Amendment] legislation," the company claimed that Standard Oil's pipelines were in effect "common carriers east of the Mississippi."[7] In an informal sense, this may have been so, but the general common-law remedies available for the use of common carrier pipelines did not free independent oil producers from Standard Oil's control over the refineries at the other end of its ostensibly common carrier lines. Whether common carriage, even coupled with expanded rate-making authority, constituted a remedy for independent producers in an industry otherwise dominated by a single company would soon be debated in the Senate.

The ineffectiveness of state regulation on interstate common carriers was most evident in the case of oil pipelines in Kansas. Oil production in Kansas soared in 1903 and 1904. To reach markets east of Kansas, oil producers were obliged to use the lines of Prairie Oil & Gas, a unit of Standard Oil. Prairie bought the oil to ship east and, with a glut of local production, reduced production-area payments from $1.30 a barrel in late 1903 to $0.80 in 1904, while the prices for refined products remained level. Prairie also instituted new prices based on the specific gravity of the crude oil, and in early 1905, the company suspended the construction of new pipeline and storage facilities. Local producers saw this as an unfair exercise of market power by Standard Oil. Prompted by Prairie's strategy, the Kansas legislature passed bills authorizing construction of a state-owned refinery. It also declared oil pipelines to be common carriers, established regulations for railroad and pipeline rates, and banned price discrimination in the marketing of petroleum products in Kansas. These regulations proved ineffective, however, in aiding oil producers seeking access to markets east of Kansas.[8]

The Garfield Investigation and Congressional Action against Standard Oil

The events leading to the federal regulation of oil pipelines by the ICC in 1906 are a fascinating juxtaposition of public opinion, politics, and the economics of pipelines. It is certainly true that the major impetus for federal regulation in the oil industry, to which the regulation of pipelines was closely related, was the obvious exploitation of rail-based oil transportation by the Standard Oil Company and its railroad collaborators. Manifested in secret railroad deals and rebates, Standard Oil's market manipulation drove down prices in the producing fields in regions like Kansas and up in consuming markets like New England. When dealing with pipelines, however, Congress faced the necessity of regulating a relatively young industry with great demands for capital to connect not with growing local markets, as the railroads did, but with oil fields that would inevitably decline in productivity and profitability. In an era when capital markets were still young and when vertical integration for capital-intensive industries (particularly oil) was a way of life in the United States, Congress would not take drastic actions with respect to either oil or gas pipelines.[9]

Responding to the frustration of Kansas oil producers, Congress in 1905 asked for a federal investigation into the price disparity between crude oil and refined products in Kansas, and whether discriminatory or illegal rail and pipeline practices were a cause. The study was performed by the newly created Bureau of Corporations, a forerunner to the Federal Trade Commission, under the direction of James R. Garfield.[10] The "Garfield report," more than five hundred pages of data and maps, is an impressive study of how the nation's rail and pipeline systems operated and how they and the transportation of petroleum were controlled by Standard Oil. The report was unambiguous in its finding that, by the early twentieth century, Standard Oil had absorbed almost all the major American oil pipelines "by means of unfair competitive methods during years of fierce industrial strife."[11] In detailed investigations, complete with foldout facsimiles of secret invoices from some of the "special agreements," the Garfield report systematically documented how Standard Oil successfully worked to monopolize the transport of oil through pipelines and rail shipments. This was at precisely the time that public opinion turned strongly against Standard Oil with the exposé of Ida Tarbell, who pioneered the field of investigative journalism in her well-documented analysis of the business methods of that company and John D. Rockefeller.[12]

Responding to the charges in the Garfield report and public pressure, Congress passed the Hepburn Amendment in 1906, adding oil pipelines to the federal regulation of railroads. The key elements of that new legislation gave the ICC the power to investigate and set maximum "just and reasonable" rates on its own initiative.[13] Given the Garfield report and the actions of Standard Oil, the inclusion of oil pipelines in the amendment was to be expected. The exclusion of gas pipelines, however, was the result of the unflagging efforts of the junior senator from Ohio, Joseph P. Foraker. This exclusion was vitally important in the ultimate institutional and competitive development of the US gas pipeline industry—as subsequent chapters will show.

Distracted by more visible and pressing railroad problems, without the accounting, licensing, or administrative tools to deal with the relatively infant oil pipeline industry, and facing a still-integrated Standard Oil Company willing to challenge its every move, the ICC spent its time on other matters. No effective regulatory control of oil pipelines in the United States happened until the 1940s, and even then no effective regulation of this industry in the modern sense happened until after Congress disbanded the ICC in 1978, handing jurisdiction over oil pipelines to the Federal Energy Regulatory Commission (FERC).

The Gas Pipeline Regulations of the 1930s

Having narrowly avoided the specter of federal regulation at the hands of the ICC, over the next thirty years interstate gas pipelines developed without any systematic form of federal price or licensing regulation. The lack of federal regulation for the interstate pipeline transport business could have been predicted to cause problems, both for gas producers in remote locations such as Kansas and for states trying to use their developing legislatively granted powers to regulate gas prices for their states' consumers.

Gas Pipeline Industry Development Since 1906

Prior to 1906, the gas pipeline industry was small and limited to the regions adjacent to known gas fields. Standard Oil used the fuel for its oil pipeline pumping stations and found limited uses in towns within 100 miles of the oil and gas fields. Pipeline technology itself limited the size of the pipeline market; the first sizable oil and gas pipelines were made of cast iron, which was brittle and unreliable for pipelines of any length. By the end of the nineteenth century, steel had replaced iron in gas pipeline construction. While steel was sturdier than cast iron, the early steel pipelines, which were rolled from flat sections, had seams that could not withstand high pressures. The largest and longest pipeline at the time was an eight-inch, 120-mile gas pipeline that ran from the gas field in central Indiana to Chicago. The introduction of oxyacetylene welding in 1911 and electric arc welding in 1922 advanced the state of seam sealing and pipe welding to the point where long-distance gas pipelines were now feasible.[14]

With such advances in technology and a growing US economy in the 1920s, the gas pipeline system grew rapidly. It was during this time that the long-distance transport of gas from the Hugoton-Panhandle basin in Kansas/Oklahoma/Texas to markets in the Midwest became possible. Figure 3 shows the major gas-producing basins and gas pipelines in 1930.

The Unsuccessful Struggle for State Regulatory Control

During the gas pipeline boom of the 1920s and early 1930s, before the Great Depression temporarily halted all gas pipeline construction, state regulators tried repeatedly to exercise control over the gas prices charged by local distribution companies. Local manufactured gas distributors had existed for almost one hundred years, but these companies had increasingly become integrated, either by contract or consolidation, into the interstate natural gas pipeline business. The charges for gas delivered to the "city gate" stations of the local distributors increasingly became a function of the gas and pipeline fees charged by companies outside state jurisdiction. A legal contest ensued over whether state or federal regulators were empowered to oversee the charges paid by consumers in their jurisdictions.[15]

Starting in 1910, the Supreme Court used a series of cases to clarify

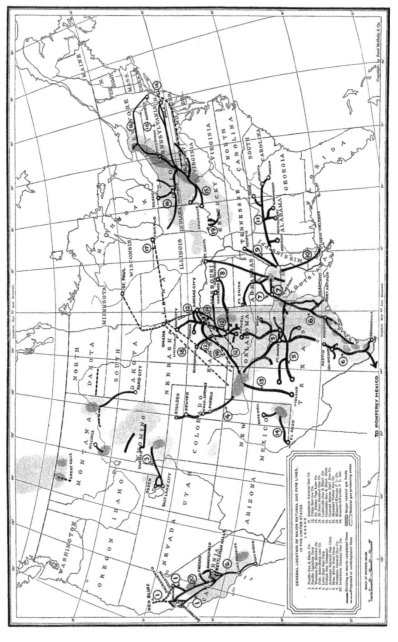

FIGURE 3. Major US gas pipelines in 1930, before the Natural Gas Act (Reprinted from J. C. Youngberg, *Natural Gas, America's Fastest Growing Industry* [San Francisco: Schwabacher-Frey, 1930], 58)

and reaffirm the necessary role of Congress, rather than the states, in regulating interstate gas pipelines. The first case involved the question of whether Oklahoma could effectively limit the transportation of gas to points outside the state. The Supreme Court ruled that such a restriction placed an undue burden on interstate commerce.[16] More telling was a 1924 case in which the Supreme Court struck down an order issued by the Kansas Corporation Commission that fixed city gate rates charged by the Cities Service system. The court preferred the "uniformity of government non-action" to an inconsistent patchwork of state regulations of interstate pipelines.[17] By the mid-1920s, the law was clear that gas transported across state lines constituted interstate commerce under the US Constitution, and Congress was the only body that could write regulations for the industry.[18]

Despite the clear line the Supreme Court had drawn delineating state and federal responsibilities for setting gas rates, state regulators could still try to pursue gas pipelines for local rate-making purposes. With the expansion of multistate holding companies in the 1920s and 1930s, state regulators increasingly saw the need to investigate transactions among affiliates of the companies they regulated. In a 1932 case, the Kansas Corporation Commission attempted to adjudicate the wholesale gas prices paid by the Western Distributing Company, which served several towns in Kansas, to its extensively integrated holding company parent, Cities Service Gas Company.

In the legal challenge to this state investigation of interstate gas pipelines, the Supreme Court faced a situation in which a state-regulated distributor purchased gas and transportation services from affiliated producers and pipeline companies owned by one of the nation's largest holding companies. The Supreme Court could hardly deny state regulators the ability to investigate the reasonableness of the upstream charges that the local monopoly utility was attempting to pass along to the state's consumers. The Supreme Court upheld the decision of the Kansas Commission, ruling that it was reasonable for such regulators to demand a fair showing that the affiliated charges were reasonable and reflected the true value of the services rendered.[19]

After that 1932 Supreme Court decision, a number of states began investigating the reasonableness of city gate gas rates (including gas and transportation charges) for holding company subsidiaries.[20] These investigations often ran concurrently with the broader movement to investigate the holding companies themselves, an inquiry that first began in

1928, when the Senate asked the Federal Trade Commission (FTC) to investigate and report on the condition of existing public utility holding companies. Business historian Christopher Castaneda called the holding company probe "one of the most widely publicized and extensive investigations of all time."[21] That FTC investigation ultimately produced the Natural Gas Act of 1938.

The Natural Gas Act of 1938

When Congress regulated interstate oil pipelines in 1906, the effort was an afterthought to a bill designed to regulate railroad rates as an attack on the exercise of market power by Standard Oil. When Congress moved to regulate interstate gas pipelines in 1938, it was filling a regulatory vacuum that had developed with the rapid expansion of the industry in the 1920s and early 1930s. Because the statute was purpose built for an existing industry, Congress had constituencies to satisfy, and it did so in three ways. First, to satisfy the states, Congress had to identify explicitly that the law did not apply to local distribution companies. Second, to satisfy existing pipeline users (principally local distributors), Congress included provisions requiring that pipeline companies' commitments to secure supplies for particular customers would not be interrupted, thereby avoiding anything resembling the common carriage requirements of the Interstate Commerce Act applied to railroads and oil pipelines. Third, incumbent pipelines themselves (and the states in which they were based), represented a powerful constituency, and Congress included licensing provisions limiting entry in all markets already served by existing pipelines.

Apart from satisfying these constituencies, Congress enjoyed two institutional foundations for creating pipeline legislation in 1938 that did not exist in 1906. First, it inherited an industry that had practically no vertical integration with gas distribution. In what constituted a severe crackdown on the industry, Congress in 1935 had given the Securities and Exchange Commission unprecedented powers to dissolve the multistate holding companies that had linked gas pipelines to local gas distribution companies. Thus, the Natural Gas Act could deal with a gas pipeline industry in effective isolation. Second, Congress drew upon more than twenty years of intensive development in utility regulation at the state level and in the courts, including explicit regulatory accounting systems, regulation of depreciation practices, administrative procedures for hearings and appeals, and so on.[22]

Regulating Gas Pipelines in Europe

Europe today has a gas pipeline system as extensive as that in North America, supplying about 25 percent of Europe's energy needs.[23] Built primarily after World War II, the system supplies Europe with gas from diverse internal sources as well as major gas fields in the North Sea, Algeria, and Russia. Maps of the gas flows in Europe show major sources coming from all directions, with the largest single flow heading west from the vast gas fields in Russia. The same type of maps for the United States would show a similarly diverse range of gas sources, with the dominant flow heading north and east from the Gulf of Mexico. Viewed from outer space, these two great gas pipeline systems would appear to have a lot in common.

But the visual similarity between the US and European gas pipeline systems is misleading. It masks great institutional and market differences. The gas production sector in the United States is highly competitive, where the top twenty producers have traditionally accounted for no more than half the market. In Europe, Russia serves close to 30 percent of the market, and the top five producers account for more than 85 percent.[24] In the United States, gas trades under contracts indexed to gas spot prices at the various physical points (called "hubs") where independent pipelines come together. The country has a competitive gas market in which prices are formed in large and highly liquid spot markets, independent from the similarly liquid spot markets in oil. Europe has no such freely competitive gas market. In the United States, gas and gas futures contracts are traded at the New York Mercantile Exchange (NYMEX), a testament to the fact that gas has become as fungible and readily transported as grain, beef, or any other commodity. The EU pipeline system also has a few gas-trading hubs and the new European Energy Exchange (EEX), but both the hubs and the EEX handle tiny volumes in forward trades in particular compared to their US counterparts.[25]

The contrast between the gas markets in the United States and Europe extends to their pipelines. In Europe, all of the major transport pipelines were built by governments or state-owned companies, and they were all integrated with local distributors (many still are).[26] The European Union has no "commerce clause" like that in the US Constitution that would reserve for the EU authorities—as opposed to national regulators—jurisdiction over trade among the EU's member states. As a result, Europe regulates pipelines via a complex patchwork of EU and national regulations.[27] Europe has no system of tradable legal entitlements in gas transport capacity. Some minor short-term swapping of pipeline capacity does occur, but it is at the discretion of the pipeline companies and is not analogous to the trade in capacity rights that characterizes the modern market for gas transport capacity in the United States. In Europe, pipelines sometimes publish measures of available capacity, but they are not uniquely defined and pipeline companies retain control over the figures. Names of shippers are kept secret by the pipeline companies, and information about shipping gas across the continent is unavailable to open scrutiny, imposing a heavy logistical and information burden on European shippers that would independently attempt to ship gas supplies across Europe. Most individual national regulators in the European Union have pushed for "retail choice" among gas suppliers by gas customers, but it is largely an ineffective choice given the lack of transparency on the EU gas pipeline system and the fact that most gas supplies to Europe flow under long-term contracts between producing countries and Europe's various nationally dominant, vertically integrated pipeline companies.

The United Kingdom: Structural Consequences of a Hasty Privatization

The 1986 sale of state-owned British Gas to private investors was the first in a worldwide wave of gas industry privatizations. The sale created two sources of conflict with which the UK regulators and British Gas (and its successors) have wrestled since. First, the government privatized a vertically integrated monopoly for reasons owing more to personalities and political exigency than as the result of an informed policy choice. Promoting gas supply competition in the context of a *private* national gas pipeline and sales monopoly subsequently proved very difficult to do. Second, the speed of the privatization process during 1986 placed British Gas into the hands of private investors well before the United Kingdom developed any procedural or accounting basis for regulating private utilities. Much as the United States struggled to create such institutions early in the twentieth century, the United Kingdom has struggled to create the same types of institutions since 1986. At the time of the publication of this book, the privatization effort is still a work in progress. There have been some advances in accounting and administrative procedures (including reversals of regulatory decisions on appeal to the courts), and there has also been a growth of competitiveness in gas commodity trading. But the gas pipeline system—the National Transmission System (NTS)—remains in the hands of a countrywide monopoly with an obscure transport pricing regime based on hypothetical gas movements to and from a central "balancing point." The pipeline pricing regime has no ability to convey meaningful pricing signals, prevents the growth of a market in transport capacity, and effectively bars any competitive challenge to the dominant incumbent pipeline company.

The NTS is a rather small island pipeline system compared to the other pipeline systems studied here. UK regulators have thus far preferred to accept their pipeline monopoly as a vehicle for the competitive trade in gas, a policy that dealt with their greatest privatization problem—the "functional" vertical integration (i.e., transport and gas sales) of the privatized British Gas. The "structural" vertical integration of British Gas (i.e., transport and distribution), or any prospect for rivalry in pipeline transport, has never appeared to be a policy imperative in the United Kingdom.

It is perhaps unfair to look upon the United Kingdom's pipeline pricing system as having any particular relevance to more extensive pipeline systems. And yet, the hypothetical pipeline pricing model—called "entry/exit"—has been widely copied in continental Europe and even enshrined in EU legislation. In that respect, the origin of this pricing system, and its limitations in efficient pipeline pricing in the United Kingdom, is a useful area of analysis for the subsequent sections of this book that deal with the transactions structure and institutions supporting pipeline transport markets.

THE ORIGIN OF THE PRIVATIZATION OF A VERTICALLY INTEGRATED MONOPOLY.   British economists have written extensively about the history of British Gas leading up to its privatization.[28] Prime Minister Margaret Thatcher and Nigel Lawson, then chancellor of the exchequer, wanted to separate British Gas vertically along supply and transport lines, as well as to accomplish a regional separation of distribution companies.[29] But Sir Denis Rooke, chairman of the publicly owned company, who was known as the "Lion of British Gas," wanted to privatize a vertically in-

tegrated company, combining the inherently natural monopolistic distribution operations with the potentially competitive gas transport or gas-
retailing functions. Rooke felt that the company should be kept whole
to enable it to "compete around the world" with other globally active,
vertically integrated gas companies like Gaz de France and Ruhrgas.[30]
Thatcher and Lawson yielded to Rooke's wishes, but the way that British
Gas was privatized is now widely viewed as a mistake.[31]

Admittedly, it is easier to criticize the structure of the UK gas privatization effort in hindsight. Prior to privatization, the United Kingdom
did not appear to give any serious consideration to implementing realistic open access to the pipeline system for independent gas sellers.[32] By
1986, the United States had not yet succeeded in implementing open access to promote gas supply or transport competition, including competition in alternate modes of gas delivery to end users for both transport
options and storage. Nevertheless, even from the perspective of allowing
competitive purchases for large gas consumers in the United Kingdom,
the privatization was a missed opportunity to create a more competitive
and efficient industry structure.

The privatized British Gas was a vertically integrated gas supply, storage, transport, and distribution company. There were no independent
distributors that could play the role of pressure groups on behalf of connected customers. As a result, the UK government focused instead on
alternative "shipper-traders," elsewhere known as "marketers," as the
source of gas supply competition on the monopoly gas transport/distri-
bution system.

ENTRY/EXIT AND ABSTRACTING FROM TRANSPORT   Following a 1992 report by the Office of Fair Trading (OFT) on monopolistic practices at
British Gas, the company committed to a system that would facilitate
the entry of competitive shipper-traders.[33] To do so, in 1992 British Gas
conceived a plan to formulate the pipeline transport system like a large
vessel, with one charge to "enter" the vessel and another separate charge
to "exit." The summation of the two nonlinked charges constituted the
transport tariff, like a trucking firm with a "pickup" and "drop-off"
charge within a city, with no explicit charge for hauling between the
pickup and drop-off points. The problem with pickup and drop-off
transport charges on a highly directional gas pipeline system is obvious:
the greatest expense in pipeline transport resides in the capital-intensive
facilities for hauling gas from point to point, not the pickup (entry) and

drop-off (exit) charges. Concealing the distance between the gas source and the customer obscured any reasonable pricing signals that a transport tariff system could otherwise convey.

Under the entry/exit regime, all gas on the British Gas system was assumed to travel to a central, hypothetical "national balancing point" (NBP), from which it was subsequently shipped back to its respective users. Creating a transactions and balancing regime for this entry/exit model proved difficult and costly. As of 1996, four years after its start, the commercial and logistic side of the regime, called the "Network Code," had cost British Gas alone in excess of £180 million and was viewed as difficult, obstructive, and unfair by gas users and shippers.[34] In late 1995, the company held seminars with its various customers and shipper-traders regarding the need to develop clearer principles and methods in setting its transportation tariffs.[35]

At the time, British Gas described the single, unified NBP marketplace as an interim measure designed to speed the entry of new gas shipper-traders into the market. In 1995, British Gas began considering ways to tie pipeline pricing to actual pipeline use in an effort to create a permanent pricing regime. Ultimately, however, the new proposal made little headway with either the Office of Gas Supply (Ofgas; or its successor, the Office of Gas and Electricity Markets [Ofgem]) or British Gas itself. British Gas probably perceived that its profitability was not linked to improving the tariff structure for transport. In any event, entry/exit was an effective barrier to entry in the market for gas transport. The proposal to make a more realistic regulated transport regime was dropped in early 1996 after some discussion, and British Gas retained the single NBP.[36]

Regulatory Difficulties for EU Gas Pipelines on the Continent

The European gas pipeline system consists of a complicated mix of publicly and privately owned and operated pipelines subject to various forms of regulation by national regulatory bodies. As one of the most obvious backbones of inter-European trade, the European Union (via its three fundamental institutions—Commission, Council, and Parliament) has made three attempts to create legislation to promote a common regulatory framework, recognizing that important regulatory functions for each member state's pipelines lie within the jurisdiction of their own regulators. The motivation of the European Union to promote such a common framework is reflected in Title XII of the treaty establishing the European Community.[37]

The first attempt at EU gas regulation was in 1998.[38] That directive was very general and high-level. It called for transparency, access to upstream pipeline systems, a competitive market in gas, and nondiscrimination among suppliers, pipelines, and gas users with regard to rights and obligations. At the same time, that first directive recognized two constraints on any full EU regulatory activity. The first was the continued ability for pipelines to "preserve the confidentiality of commercially sensitive information," which effectively meant that the European Union was barred from compelling the provision of data, or at least using those data for the purpose of regulation (Article 12). The second constraint involved the existing long-term gas purchase contracts. If the provisions for upstream access cause "serious economic and financial difficulties" to a pipeline holding such contracts, then the pipeline could be excluded from those open access provisions (Article 25). The directive called for no mandatory accounting or provision of operational information for interconnected pipelines.

In 2003, the European Union made its second attempt to regulate its continental pipeline system.[39] This second directive specified the implementation of a system of third-party access (TPA) to the transmission and distribution system, "applicable to all eligible customers . . . applied objectively and without discrimination between customers" (Article 18.1). Mindful of the different legal basis for most of Europe and the common law, this provision has similarities to a passage on common carriage in the Interstate Commerce Act in the United States.[40] The second directive prohibited the vertical integration between transport pipelines and gas distributors, "at least in its legal form" (Article 9). That statement provided an effective loophole by allowing integrated pipelines to create a paper corporate separation between transport and distribution. With respect to accounting and the provision of information, the second directive sided again with pipeline companies' desire to preserve the confidentiality of their financial and operational information (Article 10.1).

In 2007, the second directive of 2003 came under strong criticism from the EU director general for competition (DG comp).[41] Among other things, that critical report said that the continuing level of vertical integration had negative repercussions on the functioning of the market, which "constitutes a major obstacle to new entry and also threatens security of supply."[42] The DG comp report also complained about the lack of unity between the European Union and the various national regulatory authorities.[43] In general, the report highlighted a growing frustration with the loopholes in the 2003 directive. Particularly, the DG comp report highlighted the inability of the regulations to deal with the secretive, vertically integrated, and ineffectually regulated gas pipeline companies of the European Union.

In response to this frustration, the European Union proposed a third directive in 2007, which it adopted in 2009.[44] The discussion associated with the proposed third directive acknowledges that the existing unbundling (i.e., vertical separation of pipeline companies) provisions were inadequate, that the various national regulators in the member states needed strengthening and coordination among one another, and that the gas pipeline companies needed to cooperate better to facilitate seamless gas trade among the member states across the continent. Loopholes still existed in the proposed third directive, however, particularly related to the vertical separation of pipeline companies (from producers and distributors) and the provision of reliable information on costs and available transport capacity. Vertical separation could be avoided by giving pipelines "a choice between ownership unbundling and . . . setting up system operators which are independent from supply and production interests."[45] Regarding transparency and accounting, the proposed directive did not call for a standard system of regulatory accounts, and it preserved the ability for pipelines to retain what it generally labeled confidential pipeline information (Article 16 in the final directive).

Regulating the European Union's pipelines to provide for the transparent and competitive movement of gas from buyers to sellers across the continent is a major challenge. The three directives and the criticism of the lack of competitiveness of gas in the European Union attest to those difficulties. Indeed, one of the principal criticisms of the developing regulations by the DG comp in 2007 was that the continental regulation of pipelines lacked teeth in mandating transparency, in requiring vertical separation, in the unity of national regulatory rules, and in the powers of the EU regulatory body in general. The agency implicitly recognized that effective European Union pipeline regulation cannot be achieved with loopholes in those areas and without the broad institutional and political foundation needed to close them. As such, gas pipeline transport regulation across the European Union remains ineffective with the third package. Much effort is being spent on setting up independent system operators to manage complicated short-term cross-border capacity arrangements. More effort still is being spent on methods for partial cooperation among national regulators. But owing to those loopholes in the third package—among other institutional factors that would form the basis for effective European Union–wide pipeline regulation—there has been no real progress toward a regulatory regime that would break the market dominance across the European Union of national monopolies with respect to either pipeline transport or gas supply.

Privatizations and Structural Problems in the Southern Hemisphere

The Southern Hemisphere contains two large-scale pipeline systems—in Australia and Argentina.[46] There are vivid contrasts between these two countries in the structure, regulation, and investment health of their respective pipeline industries. Argentina's gas pipeline system is considerably older, dating back to the late 1940s. Australia's, begun in the 1960s, has expanded much more rapidly since the turn of the twenty-first century to provide gas to new markets. Both countries privatized their pipeline systems in the 1990s. But where Argentina usefully restructured its gas sector to create both independent rival gas pipelines and a commercial regime of transparent, regulated contract-based services, Australia conspicuously did neither. Despite creating a usefully competitive pipeline sector, however, Argentina also created a regulatory agency that demonstrated an ability to act peremptorily and/or politically—which perhaps is not surprising in a country with no modern history of utility regulation, regulated accounting standards, administrative procedures, or a reliable judicial system. For its part, despite missing the chance to promote pipeline transport rivalry, Australia has demonstrated considerable common-law due process in its application of regulation, including periodic high court reviews and reversals of regulatory opinions.

Quite apart from its regulatory challenges, the Argentine federal government ultimately wrecked the credit of its regulated pipeline industry in 2002 by unilaterally removing currency protections from gas pipeline concession contracts before devaluing its peso, which drove those companies rapidly into default. Australia can attract new investor capital for major pipeline projects; Argentina cannot, which continues to cause gas supply shortages in that country.

Australia: Pursuing Competition in an Underdeveloped,
Government-Owned Gas Pipeline System

Despite its size, Australia is still an island economy whose natural gas industry developed relatively late, owing partly to the inconvenient locations of its major sources of gas. Australia's gas fields are either in the sparsely populated interior, on the distant northwest coast, or offshore of the southeast coast in the Bass Strait. Starting in the 1990s, Australia transformed its publicly owned gas pipeline sector. A prominent 1993 government commission studying the country's slipping international competitiveness recommended the pursuit of competition at all levels of the economy, including gas pipelines.[47] That study prompted the privatization of the single federally owned pipeline serving New South Wales and Victoria's vertically integrated gas pipeline and distribution operations. Since those privatizations, Australia has developed four new pipeline links that allow the two dominant gas suppliers to sell in areas that had previously been unserved or served only by others. But the gas pipeline industry in Australia has become both highly concentrated and largely unregulated because of missteps in the initial privatizations and the rejection by Australia's competition authority of an attempt to extend effective regulatory control to the new pipelines.

By the early 1990s, the eastern Australian gas industry consisted of a handful of local gas distribution companies that served the capital cities of the states of Victoria (Melbourne), New South Wales (Sydney), South Australia (Adelaide), and Queensland (Brisbane), with some gas distribution service to outlying parts of those metropolitan areas.[48] At the time, Australia had two of the largest gas distribution companies in the world—the state-owned Gas and Fuel Company of Victoria (GFCV) in Melbourne and the investor-owned Australian Gas Light (AGL) in Sydney. There were smaller, investor-owned gas distributors operating in Adelaide (the South Australian Gas Company, a subsidiary of SAGASCO Holdings Ltd.) and Brisbane (Allgas Energy Ltd. and Gas Corporation of Queensland Ltd.).

The larger gas companies of Melbourne and Sydney were over one hundred years old, like many other of the world's major gas distributors, and were converted to natural gas only in the 1960s and 1970s. GFCV converted to natural gas in the late 1960s with the discovery of natural gas in the Bass Strait and the construction of a 250-kilometer link from the producers' processing plant to the city of Melbourne. AGL converted to natural gas in 1976 with the construction of the 1,300-kilometer, federal-government-owned Moomba-Sydney pipeline from the Cooper basin in the interior part of South Australia. The Cooper basin began to supply Adelaide with gas in 1969 through the Moomba-Adelaide pipeline, and Brisbane first received gas during that same year with the construction of the Roma-Brisbane pipeline.[49] Both pipelines were owned and operated by state-owned agencies and were part of a system of state-owned pipelines running from one gas source to each of the four state capitals in eastern Australia. With the exception of the Commonwealth-owned Moomba-Sydney pipeline, no gas crossed state lines. This reflected the strong internal interests of the various state governments, a comparatively weak federal government, and traditional objections to shipping one state's resources to another.[50]

A NATION'S PLEA FOR COMPETITIVENESS. The Australian government's ownership of various gas pipelines as well as GFCV was consistent with governmental acquisition of infrastructure businesses in British Commonwealth countries after World War II. But, as in New Zealand, slow economic growth in the 1970s and 1980s following the structural shock of the OPEC oil embargo prompted a reexamination of the wisdom of state-sanctioned or state-owned monopolies. The ability of the Commonwealth authority to intervene against monopolies was weak, and the primary Australian equivalent to the US Sherman Antitrust Act of 1890, the Trade Practices Act of 1965, included significant exemptions. Prime Minister Bob Hawke observed in 1991 that the patchwork of trade regulations among the states coupled with the practices of state-owned enterprises, including pipelines, hurt competition and consumers.[51]

Hawke began a campaign to establish a "new federalism" that would result in greater cooperation between Commonwealth, state, and local governments on issues from environmental protection to monetary policy. The landmark Hilmer report of 1993 was an investigation into Australian competition policy based on these new federalism ideals.[52] The report cited the poor productivity performance of Australia's infrastructure industries, including the gas industry, as one of the factors that had kept Australia's per capita growth rate below that of other Organization for Economic Cooperation and Development (OECD) countries. The report recommended sweeping competitive reforms, including the restructuring of public monopolies, arrangements for third-party access to energy transport systems, a more comprehensive application of price regulation, and rules for competitive conduct in all infrastructure industries.

The Hilmer report created a comprehensive road map for boosting Australia's competitiveness in many sectors, and the Hilmer committee targeted various state-owned utility enterprises in particular.[53] The committee recognized that simply privatizing state-owned businesses was not a sufficient remedy for the lack of competition, saying, "The noncompetitive habits developed through decades of operation in a tightly regulated environment run the risk of being perpetuated through private arrangements."[54] In its discussion of "public monopolies," the Hilmer report emphasized the potential damage to markets arising from the misuse of the ability to bar access to assets like pipelines. It argued in favor of full structural separation in the ownership of those assets.[55]

The Hilmer report was received enthusiastically by the various state governments and was signed by each of Australia's state premiers (as well as Prime Minister Paul Keating).[56] It prompted privatizations of the Moomba-Sydney pipeline and GFCV and the adoption of the 1997 National Third Party Access Code for gas pipelines. A decade into the twenty-first century, however, progress toward the Hilmer report's goals with respect to the gas industry has been poor. The two privatizations did not create potentially competitive pipeline businesses—the result was quite the contrary. Furthermore, the new pipeline regulation failed to deal with entry-deterring behavior by both incumbent pipelines and their regulators. Specifically, AGL and its affiliates worked to impede the Eastern Gas Pipelines (EGP) from entering the market to ship from the Bass Strait to Sydney in competition with AGL's pipeline from the more distant, and smaller, Moomba field in Australia's interior. Ultimately, the pipeline sector evolved to be concentrated in the hands of three major pipeline companies that, for the most part, have successfully avoided regulatory control.

STRUCTURAL PROBLEMS WITH TWO PRIVATIZATIONS   Australian governments privatized two pipelines in the aftermath of the Hilmer report: the 1,300-kilometer, Commonwealth-owned Moomba-Sydney pipeline in 1995 and the 250-kilometer, Victorian-owned pipeline serving the city of Melbourne from the Longford processing plant that received the gas from the Bass Strait fields in 1998.[57] From the Hilmer report's perspective of promoting competitiveness, neither privatization was a success.

The Moomba-Sydney pipeline was sold through a trade sale to AGL, the distributor in Sydney, for its high bid of A\$534 million.[58] The Hilmer committee foresaw tension between the pursuit of competitive market structures and the desire on the part of the public entities' owners to sell a protected monopoly to the highest bidder. The Hilmer committee labeled the latter choice as trading "cash for competitiveness" when privatizing government enterprises.[59] But the federal government did precisely that, selling the potentially competitive supply line to a consortium led by the gas distributor AGL. As a predictable result, the gas consortium in the closer and much larger Bass Strait fields encountered heavy resistance from AGL when it sought to enter the Sydney gas market in the late 1990s through a new 800-kilometer pipeline route up the eastern seaboard of Australia. The pipeline entrant, the A\$450-million EGP, developed by Duke Energy, had great difficulty in securing rights on AGL's trunk pipelines as a part of its project from the Bass Strait. Exasperated by AGL's evident stalling of the negotiations for capacity rights on a southern trunk pipeline that would give it access to the Sydney metropolitan area, Duke spent approximately A\$28 million to duplicate an entire segment of AGL's pipeline system, leaving the preexisting AGL pipeline underutilized.[60] Such behavior in the market for pipeline transport to Sydney has its roots in transaction cost economics (the pull of vertical integration), more traditional antitrust economic theory (raising rivals' costs), and the political influence required in order to allow such manifest inefficiencies to occur.

The EGP eventually entered service in August 2000. At the time of the EGP's construction, another pipeline link existed between the Bass Strait and Sydney that the EGP was built explicitly to bypass. The Victorian and New South Wales gas pipeline systems had already been linked by the A\$55-million Interconnect, completed two years earlier. That link has remained an ineffective route for shipping Bass Strait gas to New South Wales for two reasons. First, the majority (51 percent) owner of the Interconnect was again controlled by AGL, meaning that any distant Cooper basin volumes displaced by closer Bass Strait volumes caused a loss of revenue to the AGL-affiliated Moomba-Sydney pipeline.[61] Second, the transport arrangements instituted with the 1998 privatization of GFCV effectively prevented the Bass Strait producers from using existing links to ship gas to New South Wales.

The Victorian pipeline was privatized not as a contract transporter but as a "notional" transmission system created to support pool-based spot trading for gas to be managed by the same independent system operator that managed the pool-based spot trading for electricity in the state (VENCorp).[62] Without contracts or any other way to secure long-term commitments for gas transport, the Victorian privatization created monopoly gas transport for Melbourne, bundled with Victoria's LNG storage, managed by a state-governed bureaucracy. It was a move that effectively removed all avenues of pipeline rivalry in addition to foreclosing the use of contracts for transport on the existing pipeline system within the state or to New South Wales. Victoria's new regime was called "market carriage," and the 1997 National Third Party Access Code was quickly amended to permit such a pool-based pipeline system.[63] Despite the objections of the gas suppliers and others, the Victorian Treasury proceeded with the privatization of one transport pipeline, three gas distributors, and three separate gas traders.[64] The state continued to purchase the supplies from the Esso/BHP consortium and in turned resold the gas to these three traders (for resale to final consumers).

The Victorian mandatory short-term gas pool mechanism continues today. The price for gas has occasional needle spikes, but it always reverts to the price settled in the periodic protracted arbitrations in the contracts between the single major supplier and the three retailers. Despite its considerable complexity and cost (both direct and indirect) and the vigor with which the spot-trading agency defends its role in setting delivered gas prices in Victoria, the "market carriage" arrangement achieves nothing in terms of promoting gas competition in Victoria or the efficient use or expansion of the pipeline transport system.[65]

THE CONTINUED CONCENTRATION OF THE AUSTRALIAN GAS PRODUCTION SECTOR. The concentration of gas producers in the eastern Australian gas market is a problem. At the time of the Hilmer report, each of the four cities in eastern Australia was served by a single gas production joint venture. Santos and its joint venture partners served Sydney, Brisbane, and Adelaide, and the Esso/BHP joint venture served Melbourne. Pipeline construction has since connected the Esso/BHP consortium to Sydney, Adelaide, and Hobart (in Tasmania). These two joint ventures control most of eastern Australia's gas supplies.

In 1996, the Commonwealth energy regulator, the Australian Competition and Consumer Commission (ACCC), attempted to weaken the monopoly control over gas sales by the Cooper basin joint venture by effectively revoking a 1986 authorization for a joint contract between those producers and AGL. One of the ACCC's biggest objections to continuing with the 1986 agreement was its Clause 12, which gave the joint venture the right of first refusal for supply of additional volumes to AGL.[66] The ACCC was convinced that, by 1996, "the anticompetitive detriments of the authorized agreements outweighed their public benefits" and there were sufficient grounds to revoke the earlier authorization.[67] The Australian Competition Tribunal (ACT) disagreed, holding that the original joint agreement was "intrinsic also to the achievement of the benefit that arises from its existence and implementation."[68] That is, the ACT was thus convinced that if it took such a monopolistic provision to get the project going in the first place, the provision should be retained. As a result, however, the Cooper basin joint venture, led by Santos, continues to market gas jointly and hold a right of first refusal, effectively monopolizing the sale of gas through the Moomba-Sydney and Moomba-Adelaide pipelines.

The persistent concentration in the Australian gas-producing sector, along with the difficulty for shippers to arrange for transparent transportation between the states (a function of pipeline deregulation as well as Victoria's "market carriage"), has generally prevented the rise of an independent gas market in Australia free of the type of large-scale price negotiations (now large-scale arbitrations between buyer and seller interests) that have been part of the Australian gas industry since its beginning. In the United States, the competitiveness in the gas market depends both on underlying competition in the supply of gas and on a flexible, fully transparent, contract-based pipeline transport system. In the early part of the twenty-first century, Australia is developing new sources of gas that may start to weaken the highly concentrated production and marketing sector.

Argentina: Restructure First, Privatize Later

Argentina's Gas del Estado was created following the 1945 nationalization of Argentina's gas industry by Juan Perón's government. Before then, gas demand in Argentina was met by processing coal imported from England (and distributed by British-owned gas distributors). In 1947, looking to develop its own fuel resources, the Argentine government began constructing the country's first gas pipeline, running from the oil production center in Comodoro Rivadavia in Patagonia to the capital city of Buenos Aires.[69] The pipeline, which measured 1,700 kilometers long and ten inches in diameter, began operating in late 1949. At the time, the Comodoro Rivadavia–to–Buenos Aires pipeline was one of the longest gas pipelines in the world. Following its construction, the Argentine government implemented and pursued a policy of rapid gas service expansion. By 1988, gas met nearly 40 percent of Argentina's primary energy needs and comprised 36 percent of its final energy consumption.[70]

Along with this rapid growth came great demands for capital, which was a persistent problem for Argentina under its postwar Peronist governments. The country's shortage of credit caused it to seek foreign private capital for the development of pipelines. In 1970, Cogasco, a Dutch-led consortium, began raising external capital to build and operate the Centro-Oeste Pipeline to the newly discovered gas fields southwest of Buenos Aires, eventually to transfer ownership to Gas del Estado. The 1,125-mile Cogasco pipeline commenced operations in 1981 under a state guarantee to convert local peso revenues to hard currencies during Argentina's seemingly perpetual hyperinflation in the 1980s. But during the military conflict with the United Kingdom in 1982, when there was an extreme shortage of hard currency, the government claimed a technical breach of contract and stopped all peso payments to Cogasco, thereby mooting the convertibility issue and causing the firm to fail by the late 1980s.[71] Foreign investment in the Argentine gas sector stopped after this episode, which demonstrated the government's failure of commitment to such a pipeline enterprise.[72]

THE MARKET PRIOR TO PRIVATIZATION. Despite its long and important history in supplying Argentina's energy needs, Gas del Estado found itself unable to meet the winter peak demand by the late 1980s. Many of the industrial and power generation customers in Argentina's major markets, particularly the principal market of Buenos Aires, had developed dual-fuel capability to deal with this unreliability in winter gas supply. As a consequence, the preprivatization annual load curve showed a relatively steady load throughout the year, illustrating the ubiquitous weekend dips seen in most temperate countries, accompanied by a stepwise rise to another relatively level plateau 50 percent higher for the winter months of June through September. This fairly unusual stepwise peak pattern resulted from increases in heating demand during the height of winter that were partially offset by decreases in electricity generation and other uses due to low pressure, interruptible contracts with industry, and curtailments. The inability to serve a more typical peak load pattern of gas use (which should have looked more like a mountain than a mesa) was a major national problem. Important winter demand load in both electricity and gas went unmet.

Prior to privatization in 1992, the gas industry in Argentina was controlled by the government. The government, through Yacimientos Petroliferos Fiscales (YPF), produced or purchased gas in the field and, through Gas del Estado, transported and distributed that gas to consumers. Because of its complete ownership of the industry, the options for restructuring the gas industry prior to privatizing the system were wide open. The industry displayed an unusually good potential for competition through privatization. In contrast to some newly privatized (or soon-to-be-privatized) gas markets in locations such as Australia or New Zealand, there was no *structural* feature standing in the way of the Argentine government's creating competition in gas supply, competitive rivalry among pipeline transportation companies, and a regulated distribution sector.

PROMINENT PRIVATIZATION SUCCESS IN 1991. In the time leading up to privatization, the structural potential for gas supply competition and pipeline rivalry was high. Argentina had proven gas reserves sufficient to supply thirty years of domestic consumption at then-current levels. Gas supplies moved from these reserves in the direction of Buenos Aires from three distinct directions: the Neuquén Province to the southwest, Salta in the far north, and Santa Cruz and Tierra del Fuego in the far south. In 1989, production exceeded 800 billion cubic feet (bcf). Of this, about half came from the Neuquén fields, a third from fields in Santa Cruz and Tierra del Fuego, and the remainder from Salta. Argentina also imported about 10 percent of its gas supply from Bolivia. By 1991, Argentina had an extensive gas system, with more than 12,000 kilometers of transport pipelines.[73]

The structure of Argentina's gas supply system presented unusually attractive opportunities to promote competition in both gas and transport. In the privatization of Gas del Estado, one of the principal choices facing the government was whether to privatize single or multiple transporters. A World Bank–sponsored team advising the government at the time supported the privatization of multiple pipelines. That team pointed to the great body of evidence from the United States, where many regions of the country were served by two or more independent regulated transport companies, with the resulting competition from established rivals for the regulatory approval of licenses for the construction of facilities to meet new demand.[74]

Before privatization, several gas companies from Europe, including British Gas, had advised the Argentine government that there would be little interest among internationally active gas companies in purchasing vertically separated gas transportation and distribution franchises in Argentina. With the support of its World Bank–sponsored advisers, Argentina resisted such advice, which would essentially have duplicated the vertically integrated European industry structures. Instead, Argentina structured and sold two independent pipeline transport companies and eight independent distribution franchises.[75] The stated rationale for the breakup was to encourage competition in transport and enable the performance of various independent distributors to be gauged. There was considerable interest in the bidding, with proceeds received by the government totaling more than $3.8 billion (including cash, debt retirement, and other assumed liabilities).[76] The successful bidders included major participants from North American and European gas companies.[77]

The Argentine privatization was a prominent structural success, in one stroke accomplishing five major structural and institutional advances in creating an industry structure conducive to pipeline rivalry. First, the privatization prevented pipeline companies from owning gas volumes headed to the distributors and thus institutionalized contract-based pipeline transport. Second, it prevented cross-ownership among the gas, transport, and distribution sectors, creating parties that dealt with one another at arm's length—avoiding many of the affiliated interest problems associated with other privatizations, such as those in the United Kingdom or Australia. Third, it specified a system of transport tariffs that reflected efficient pricing parameters, such as long-term contracts reflecting distance and capacity for the transport pipelines, and peak load pricing for the distributors. Fourth, it created regulated accounting books for the privatized companies reflecting a onetime independent assessment of the capital stock that would form the basis for future regulated tariffs. Fifth, it avoided anything reminiscent of common carriage, creating the kind of contract obligations for pipeline transporters that defined transport capacity as a right, with some similarities to how transport capacity was defined in the United States.

Argentina's was the first gas industry privatization to embrace all five of these characteristics (none of which accompanied the privatization of

British Gas).[78] By eliminating the possibility of forming a vertically integrated gas monopoly, the new industry structure undoubtedly reduced
the profit potential of the remaining enterprises and the potential proceeds to the government through privatization. This was particularly
true for transport. By committing to selling two independent pipeline
transporters, the government was explicitly trading profit potential and
privatization sales value for the long-term prospect of competition for incremental growth in gas demand and the welfare of gas consumers.[79]

Pipeline Deregulation

Economic regulation of pipeline prices has generally accompanied all
major privately owned pipeline systems based on a structural assessment
that they can treat at least some of the customers they serve as monopoly
suppliers. There have been various efforts, however, to relax the strin-
gency of those regulations on competitive grounds. There are prominent
examples of such efforts at deregulation, or "light-handed regulation," in
the United States and Australia.

Light-Handed Regulation for US Oil Pipelines

Federal regulators and the US Department of Justice (DOJ) have had
institutional difficulties with regulating oil pipelines from the start.[80]
Partly in response to those frustrations, in 1984, the DOJ released a preliminary report on competition in the oil pipeline industry.[81] It then invited comments from the industry and distributed a final version of the
report in 1986.[82] Using an analysis that employed a standard market concentration measure in origin and destination markets, the DOJ did not
find any crude pipelines that presented a clear case for continued federal
regulation.[83] For all existing crude oil pipelines, it recommended deregulation. For product pipelines, it recommended a case-by-case examination to determine if respective origin and destination markets were sufficiently competitive to make a case for deregulation. Largely because
of this study, crude oil pipeline prices have transitioned to a variant of a
"price cap" regime in which prices are allowed to increase vis-à-vis the
general price index plus an ad hoc adjustment. The case-by-case examination of products pipelines has led to the deregulation of more than a
dozen.

RISING REAL PRICE CAPS FOR US CRUDE OIL LINES.   In 1992, having observed the difficulty the FERC was having in setting oil pipeline rates using the vague regulatory practices it had inherited from the ICC in 1978, industry representatives turned to Congress for guidance in defining streamlined rate-making methods for oil pipelines. The industry's move was prompted largely by the FERC's administrative challenge of dealing with a complex industry of vertically integrated joint ventures, where the pipelines themselves held no federal certificates (the Hepburn Amendment did not require them) and did not need to conform to the Uniform System of Accounts (which was created only later for gas pipeline regulation), and where many oil pipeline rates had never been litigated to a regulatory judgment at all.

Congress took up the issue of streamlining oil pipeline rates and settling the lawfulness of the obscure ICC rate-making procedures. In its Energy Policy of Act of 1992, Congress made revisions to the Interstate Commerce Act to require that the FERC issue a final rule for a "simplified and generally applicable ratemaking methodology" for oil pipelines within one year. It also directed that "any rate in effect for the 365 day period ending on the date of enactment of this Act shall be deemed to be just and reasonable."[84] Congress evidently intended to set the level of outstanding rates and simplify the costs of rate making for the oil pipeline industry going forward.

In October 1993, after finding that an indexing method similar to the price cap regulations adopted in many other jurisdictions would comply with Congress's desires, the FERC issued an order that specified an alternative rate-making policy.[85] The FERC sought a rate-making method applicable to the diverse array of oil pipelines, some of which had operated for decades without having their rates judged by any federal agency. The FERC eventually adopted a recommendation from Alfred Kahn's testimony on behalf of the Association of Oil Pipelines (AOPL). Kahn proposed to index the existing rates according to a five-year price cap formula of the producer price index (PPI) minus an ad hoc 1 percent, or "PPI − 1" for the period July 1, 1996, through June 30, 2001.[86] For 2001–6, the FERC attempted to keep the "PPI − 1" index in use by making a small switch in methodology, but the US Court of Appeals determined that it failed to justify the shift in methodology. The FERC then reverted to the Kahn method (this time supported by other witnesses), which, with the updated pipeline accounting data, became "PPI − 0" percent for the 2001–6 period. For the period 2006–11, the FERC adopted the same methodology with further updated pipeline accounting data, yielding a "PPI + 1.3" percent price cap. For the 2011–16 period, based again on evidence provided by the AOPL, the FERC boosted the cap further to "PPI + 2.65" percent.[87]

Those familiar with the application of price cap regulation around the world would see that the latest FERC action granted the equivalent of an *X-factor* of negative 2.65 percent (meaning that oil pipelines are held to be slipping by 2.65 percent in productivity, year on year, compared to the rest of the economy, necessitating a cumulative compounded five-year real price increase above inflation of 14.0 percent to compensate). Examining the record, this seemingly generous outcome for the pipelines for their fourth review period, compared to the widespread experience of other regulated firms under such price cap plans, was likely the result of ineffective opposition from pipeline shippers. In any event, the use of reported accounting data only, rather than a genuine total factor productivity (TFP) study, would not chart productivity effectively over time. To the extent that the FERC permits such regulated price ceilings to rise substantially in real terms with respect to inflation, for an essentially static technology with well-known costs, it might be held to be consistent with the agency's general desire to become less involved in the regulation of what it feels to be a generally competitive activity (consistent with the DOJ's 1986 recommendations).[88] Then again, it may merely reflect a failure of collective action on the part of oil pipeline shippers—a complex issue given the widespread vertical integration in the industry.

MARKET-BASED RATES FOR US OIL PRODUCT PIPELINES. In the case of oil product pipelines, the DOJ recommended continued regulation for five lines and reserved judgment on another six, based on its studies of market concentration for each. It also supported the "prompt deregulation of all other oil pipelines currently subject to federal regulation."[89] Despite its strong support for relying on competition to set rates on oil pipelines, the DOJ continued to be concerned about the prevalence of vertically integrated joint ventures between shipper-owners. It noted that the Interstate Commerce Act did not give the FERC a role in licensing oil pipelines (i.e., the regulation over entry and exit of capacity), as it had under the Natural Gas Act for gas pipelines. Thus, "the vertically integrated shipper-owners can evade existing pipeline regulation and gain monopoly profit by simply setting the capacity of the pipeline at its monopoly level."[90] Ultimately, the DOJ conceded that the regulation of vertically integrated joint ventures is very complex, as "vertical integration may make unnecessary the regulation of pipeline market power, but it can also thwart attempts at regulation."[91] Ultimately, the FERC has permitted "market-based rates" on more than a dozen oil product pipelines, but not for crude oil pipelines.

Continued Cost-Based Regulation for US Gas Pipelines

Regulating interstate gas pipelines caused considerable conflict among producers, pipelines, and gas distributors from the inauguration of the Natural Gas Act of 1938 through the 1990s. By the 1980s, however, the FERC was searching for any alternative to traditional pipeline regulation that would foster increasing competition in either gas or pipeline transport. Part of the FERC's search included a 1995 investigation into whether there were alternatives to cost-based rate making for gas pipelines.[92] As a response to this request for comments on alternative rate-making methods for gas pipelines, Koch Gateway Pipeline, an interstate pipeline running through the southeastern United States from Texas to Georgia (originally known as Southern Natural Gas Company), filed a proposal for market-based rates. That proposal served as a test case as to whether the FERC would relax cost-based rates for the Koch pipeline or any other interstate gas pipeline.

The FERC used the same kind of analysis introduced by the DOJ in its 1986 report to evaluate the market concentration of pipelines—that is, the concentration of pipelines within origin and destination markets. Koch had proposed using a five-state geographic area to assess pipeline concentration, contrary to the origin market and destination market inherent in the DOJ report. The judge hearing Koch's evidence agreed, but the full commission disagreed and overturned the judge's finding. The FERC held that only locally available pipeline alternatives with spare capacity could realistically constitute the basis for the kind of pipeline rivalry that would discipline transport prices. As the FERC had done in its cases regarding market-based rates on various oil product lines, it ruled that a proper geographic market could include only those pipelines from which a local customer could purchase transport service if Koch attempted to raise its prices. It thus held that Koch had not met its burden of proof and denied the company's request for market-based rates.[93]

Since the Koch case, no other interstate gas pipeline has petitioned for market-based rates, as the standard set by the FERC effectively precluded success. The ability of oil pipelines to achieve before the FERC what gas pipelines could not was a function of two factors. First, the geographic origin and destination market for oil pipelines was considerably wider than for gas pipelines, reflecting the role that roadway competition could play in oil—but not in gas. Second, the common carrier oil pipelines were not inherently contractually committed, as common carriage forbids such contracts. Gas pipelines, on the other hand, were built for users to whom pipelines had contractual commitments for most interstate transport capacity. Those contracts thus precluded the ability of one pipeline customer to necessarily be able to obtain firm transport service on another pipeline simply because it was nearby.[94]

De Facto Gas Pipeline Deregulation in Australia

Since the completion of the EGP from the Bass Strait north along the coast of Victoria and New South Wales to Sydney in 2000, three more significant pipeline links have been built in the eastern Australian market: a pipeline south from the Bass Strait to Tasmania, a pipeline from Victoria to Adelaide in South Australia, and further pipeline links in Queensland. In 2009, three companies controlled those pipelines.[95] The National Third-Party Access Code, which came out of the Hilmer report recommendations, dictated that Australian pipelines would be subject to access and tariff regulation. That position began to unravel in 2000, with the application by the New South Wales Ministry for Industry, Tourism, and Resources to extend regulation to the EGP. The National Competition Council (NCC) recommended in 2000 that the EGP be subject to access and tariff regulation, but this decision was later overturned by the Australian Competition Tribunal (ACT) in 2001.[96] The tribunal held that regulating the EGP would not promote competition in gas markets, particularly given the EGP's uncommitted capacity and the incentive for the pipeline to maximize its shipments. At the same time, the ACT felt that the provision of public information that regulation would require would be of little benefit in preventing discrimination among customers and would be more likely to facilitate collusion among alternative pipeline suppliers.[97]

The consequences of the ACT's decision regarding rejection of cost-based regulation for the EGP was the rejection of regulation for the other new pipeline companies, including those linking to Tasmania and the link between Victoria and Adelaide. The New South Wales minister of tourism, industry, and resources, the Honorable Ian "Chainsaw" Macfarlane,[98] deregulated on his own authority all but 27 percent of the length of the Moomba-Sydney pipeline in 2003, against the recommendation of the NCC. As a result, the only gas pipelines that continue to be regulated in Australia are the Victorian transmission system and two pipelines serving the smaller markets in Queensland.

The End of the Structural Analysis

So concludes the generally neoclassical structural analysis of the major pipelines and the potential for pipeline-to-pipeline rivalry. A focus on structural issues is of course essential in evaluating the potential for competitive rivalry in the pipeline business and the style of regulation suited to that potential. But such a perspective takes the analysis of pipelines, as inland transporters, only so far. A structural analysis is not sufficient to analyze the roots of the substantial differences between the kind of pipeline financing or regulation applied in North America and that applied almost everywhere else. Nor does a structural analysis deal with the split between the rights to pipeline transport entitlements and pipeline ownership or help to explain the source of the paradox associated with the juxtaposition of cost-based regulation for all US gas pipelines with unregulated prices for legal transport entitlements—all supporting a highly competitive gas commodity market. Analyzing those issues demands a type of economic analysis that includes the elements of the new institutional economics.

The Essential Contributions of the New Institutional Economics

Despite the way that it seems to have driven regulatory policy for more than a century, there is plainly more to the pipeline industry than the neoclassical concepts of natural monopoly, or the structural analysis of pipeline markets in chapter 4, can handle. Individual lines are not built once to connect a field, import terminal, or refinery to a market. They expand and evolve to serve changing markets and ever-declining oil and gas fields.[1] In the United States, major pipeline companies cross one another's rights-of-way and pass by large cities without providing local service on the way to more distant markets (imagine how strange it would be for a railroad line to pass close to a major city without the trains ever stopping there).[2] Major extensions of the gas transportation system in the United States seem to come in pairs rather than with a single, lowest-cost line. In Australia, two different gas pipeline companies operate lines that exist one atop the other over the same right-of-way in the southern Sydney suburbs. And the world's first large-diameter gas pipelines were not constructed for the gas industry at all but represented the conversion of oil pipelines that went on sale as the ultimate wartime surplus.[3] If pipelines are such powerful natural monopolies and costs decline with larger pipes, why do real pipelines interact in such seemingly strange and inefficient ways?

This chapter looks more closely at the concepts of the new institutional economics introduced in chapter 1. These involve transaction cost economics, the evolution of institutions of economic governance, the role of pressure groups (i.e., "collective action"), and the property rights elements that would set the stage for Coasian bargaining in point-to-point pipeline transport rights. As such, this chapter constitutes the rest of the story that only begins with the declining cost technology of pipelines and the structural status quo that were the subjects of the neoclassical economics in chapters 3 and 4.

To be sure, these are all rich fields of study that have produced great quantities of published research reflecting the diversity of approaches to investigating economic governance. This river of relevant research flows from the widening of the boundaries of economic theory inherent in the new institutional economics. Recent and valuable compilations of this varied research exist.[4] As a continuation of the prelude to analyzing continuity and change—both across countries and through time—for this industry with its peculiar characteristics of costs and asset specificity, this review is still necessarily brief.

The Cost of Transacting

Oliver Williamson popularized transaction cost economics and the limitations of traditional neoclassical production cost analysis with a new generation of economists. But perhaps the most articulate description of those limitations came from Douglass North in his 1993 Nobel Prize address: "The neoclassical result of efficient markets only obtains when it is costless to transact. . . . When it is costly to transact, then institutions matter. And it is costly to transact."[5] North's point is that neoclassical economics is concerned with the operation of markets, not with how markets develop or evolve. At the time, he admitted that there is no clean mathematical theory dealing with institutions to rival the rigorous and systematic development of economic dynamics or general equilibrium theory, only an "initial scaffolding of an analytical framework capable of increasing our understanding" of the development of markets.[6] During the 1990s, the theoretical underpinnings have been developed in both the transaction cost and property rights components to institutional economics.[7]

While merely an initial scaffolding, the new institutional economics is uniquely pertinent to the pipeline industry as an additional tool for microeconomic analysis.[8] Without it, the schism between US oil and gas pipeline markets in terms of organization, regulation, pricing, and performance would escape theoretical economic analysis.[9] How can one pipeline system in the United States be vertically integrated and dominated by joint ventures among the major oil producers, with no real prospect for the competitive entry of independents or Coasian bargaining in transport capacity, and the other be characterized by new and freewheeling competition (including forward and futures markets) in pipeline use and genuine rivalry in capacity planning and construction? The first element in the answer lies in the transaction structure of the industry.

The cost of transacting with pipelines once demanded vertical integration. Business information was sparse and long-term contracts difficult to negotiate and enforce in an era before the Securities and Exchange Commission made the provision of high-quality business and securities information commonplace.[10] In such an environment, the cost of contracting to deal with pipelines, including both the ex ante costs of negotiating and safeguarding agreements and the ex post costs of dealing with unanticipated contract problems and opportunistic behavior, was evidently larger than the cost of integrating in the oil companies' pursuit of profits.[11] Economists have held the ex post costs—particularly opportunism in the fact of changing circumstances ("self-interest seeking with guile," according to Williamson)—to be the most troublesome in vertical relationships. To understand how transaction cost economics has dealt with such opportunism, two elements in the theoretical investigation of such postcontractual opportunistic behavior appear—the presence of *appropriable specialized quasi rents* and *asset specificity* (or its equivalent term, *asset specialization*).[12]

Appropriable Specialized Quasi Rents

A transaction-specific asset realizes its full value only in relationship to a particular transaction and thus becomes less valuable if relegated to an alternative use. A commonly used, nonpipeline example is a rail line constructed to transport coal from a mine to an electricity generator. If the generator shuts down, the rail line may have no alternative use and will thus be rendered valueless except for the residual value of its rails and right-of-way. The capital invested in pipelines is similarly highly transaction-specific, as the only purpose of those lines is delivering oil or gas from petroleum producers to oil refineries or gas consumers.

These kinds of investments create possibilities for buyers to interfere opportunistically with sellers. Without a sufficiently exacting prior agreement, once a plant is dedicated to a particular purpose, the buyer can find some subtle or obvious pretext to reduce payments. The seller has nowhere to turn if its particular dedicated asset has only a small value in any alternative use. In this instance, *quasi rent* is the value that the buyer may capture—or appropriate.[13] In any transaction with dedicated assets, particularly with pipelines, quasi rents are vulnerable to opportunistic capture by their counterparties.[14] It can work both ways. *Bilateral dependency* is when both parties make specialized investments for the other. As Benjamin Klein, of the University of California–Los Angeles, and his coauthors, Robert G. Crawford and Armen A. Alchian, state in a much-cited paper, such investments pose a cost for which the two remedies are either vertical integration or long-term contracts. As assets become more relationship-specific, creating more appropriable quasi rents (increasing the possible gains from opportunistic behavior), the costs of contracting will generally increase more than the costs of vertical integration and we are more likely to observe vertical integration.[15] Pipeline systems, with their extreme asset specificity, raise the prospect of either costly transacting through contracts or vertical integration to internalize those costs.[16]

Asset Specificity and Idiosyncrasy in Pipeline Investments

The risk that quasi rent will be appropriated depends on three main things: (1) the frequency with which buyer and seller expect to interact in the future (increased frequency makes ex post opportunism less likely); (2) the level of uncertainty about future market conditions; and (3) the degree of asset specialization—also known as "investment idiosyncrasy"—required by the specific transaction.[17] Idiosyncratic investments are made to support a specific commercial relationship; by definition, they would be uneconomic outside the context of that relationship.[18] By their nature, idiosyncratic investments are transaction-specific. Fewer interactions, more uncertainty, and greater idiosyncrasy all increase the risk of quasi-rent appropriation. In addition, the greater the time or location sensitivity, the greater the risk.[19] The consensus in the economic literature is that of the three variables, the most important in raising the risk of quasi-rent appropriation is the level of investment idiosyncrasy.[20]

The petroleum industry is a particularly good example of these relationships, and Klein, Crawford, and Alchian considered it as a key example of the problem. Once all the assets are in place (the wells drilled and the pipeline and refineries constructed), the oil-producing properties and the refineries are specialized to the pipeline. These specialized producing and refining assets are therefore hostage to the pipeline owner (as John D. Rockefeller discovered in the 1870s). Oil field owners could remove the possibility of subsequent quasi-rent appropriation through shared ownership in the pipeline.[21]

The durability of pipeline assets demands that any contract that would substitute for vertical integration should be sufficiently complete and well defined to anticipate a wide variety of possible futures for the industry.[22] Practical experience with long-term contracts in the petroleum industry shows that this is hard to do. Given enough time, uncertainty looms large even in such a low-technology industry. A contract's "navigational aids" may shift or disappear. That is, markets develop, published pricing benchmarks cease to exist, seemingly settled industrial relationships overturn, and government regulations change. All of these factors test the prescience of even the most careful contract writers and make for risk and possible future litigation in a long-term contract setting. One view would be that all that is necessary is for contracts to be sufficiently written such that each party can recoup its initial investments (fulfilling any required rate of return). But to the extent that unanticipated changes occur in the industry ten or twenty years after these investments are recouped, then any subsequent rent-seeking behavior could well feed back into investing. The inability to put contingency clauses in contracts to handle such situations might prevent producers / refiners / pipeline companies from engaging in otherwise profitable ventures.

Given the durability of pipelines and their extreme idiosyncrasy, the theory points to an almost inevitable desire on the part of the industry to integrate rather than deal with the cost of contracting and opportunism. This desire creates two interrelated problems. First, vertical integration with major pipelines greatly concentrates what otherwise are potentially competitive oil- and gas-producing sectors, running into the social and legal barriers to monopolization in market economies. Second, longstanding economic governance institutions associated with transportation, not totally unrelated to the monopoly problem (e.g., common carriage), abhor the idea of exclusivity that vertically integrated pipelines may promote vis-à-vis nonintegrated shippers. We turn to those longstanding institutions next.

Common Carriage / Third-Party Access and
Pipeline Transacting

Foggy definitions are a problem for economic analysis. Perhaps the foggiest terminology, obscuring a clear view of pipeline market potential and efficient regulatory practice, involves *common carriage* (used in the United States) and *third-party access* (or *TPA*, used mostly in Europe but familiar almost everywhere). The terms are loosely held to be equivalent. Read literally, neither looks objectionable—but that is not where the problem lies. The problem is that common carriage and TPA mean something more: that discrimination among customers in prices for, or access to, pipeline service is prohibited. Such prohibitions include the writing of contractual agreements that charge similarly situated users different prices or that exclude some users from access in favor of others. These prohibitions may seem self-justified—which is the reason why legislators and many economists appear drawn to the concept without a second thought. But to anyone wrestling with financing relationship-specific transport assets like pipelines, common carriage and TPA are highly problematic for the almost ancient restrictions on efficient transacting they embody.

Common carriage is a centuries-old legal concept under the common law. It has been a problem for relationship-specific pipelines as it was for railroads many decades before. Some key solutions for the economic governance and financing of such independent (nonintegrated) transport lines caused by relationship specificity depend on the writing of specialized contracts that memorialize certain kinds of pricing and access priority. On the surface, such specialized contracts, which economic theory would otherwise hold as efficient or cost justified, may run afoul of statutes that contain vague or boilerplate common carriage language. Such was the case for the Interstate Commerce Act of 1887 (which impaired the development of competitive railroads and oil pipelines in the United States). Such is also true for the various legislative packages that have been considered by the EU Parliament to deal with Europe's gas pipelines. In sharp contrast, and by design, common carriage was never applied to gas pipelines in the United States. Those pipelines are "private carriers" without common carriage obligations. That distinction is the key to that industry's evolution to a competitive market in point-to-point pipeline transport with Coasian bargaining.

A digression on common carriage and TPA is unavoidable if an economic analysis of the different pipeline transport markets in the world is going to make sense.

The Development of Common Carriage Transport Regulation[23]

There are two essential features of modern pipeline regulation that have their origins in common carriage: licensing and tariff regulation. Licensing involves a "certificate of public convenience and necessity," which performs two functions. It grants commercial exclusivity by franchise, a very old practice stretching back to medieval Europe, and it exerts the government's power of eminent domain necessary to allow transport providers (like railroads and pipelines, but before them canals, turnpikes, and stagecoaches) to cross the countryside without having to engage in bilateral negotiations with individual property owners. Tariff regulation did not accompany the earliest common carriage franchises, due to the belief that profitability associated with a grant of monopoly franchise would be adequate compensation for transport companies given the risks involved. The regulation of tariffs began in the late nineteenth century, as the scale and complexity of the new common carriers in railroads and pipelines presented issues beyond what the normal common-law courts could effectively manage. Specialized agencies became necessary to resolve complex rate disputes, particularly in the case of railroads.

Common Carriage and the Duty to Serve

The principal defining feature of a common carrier is that it transports people or goods for hire, whenever those people or goods show up wishing to be transported. A common carrier holds itself ready to serve the general public to the limit of the facilities that it is prepared to offer. By contrast, a private carrier transports only its own goods or those of a narrowly defined clientele. At one time, common carriage generally covered most inland waterways, freight railroads, intercity passenger railroads, commercial airlines, and many automotive freight or passenger businesses. By the time pipelines entered the scene in the late nineteenth century, common carriers included major portions of the enterprises that carried on the business of transportation in the United States and the United Kingdom.

The concept of the common carrier developed in the United Kingdom in the form of the guild system, whereby certain commercial activities could be undertaken only under special grant of privilege that excluded the competition of others.[24] After the breakdown of that system and the rise of the competitive enterprises to take its place, the grants of privilege generally disappeared for most types of commerce. However, the concept of a special grant of privilege survived for businesses that transported goods or people for hire.[25] For much of the nineteenth century, communities recognized the benefit of the common carriers for transporting passengers and goods. As this appreciation grew, so did the tendency to assign responsibilities to the transport companies as part of the grant of common carriage with the accompanying protection from competition. Some of these responsibilities dealt with the risk of the transport mode and the prospect of losses and harm to passengers. Others dealt with the general need to assure the public that charges were fair. To the extent that common carriers wished to engage in the business, they were expected to serve all comers equally, charge reasonable prices, and bear responsibility for the safe delivery of the goods or persons committed to their care.

The duty to serve is perhaps the most elemental of the common carrier's obligations. Stuart Daggett, of the University of California–Berkeley, an early twentieth-century transportation economist, addressed the seemingly obvious question raised by this legal obligation: "Why, one may ask, should a carrier ever desire to refuse to serve?" The answer lies in the relationship between transport companies and the businesses they serve or, as Daggett said euphemistically, "some angle of business policy."[26] In many businesses, there is an affiliate interest between the source of materials or people and the place of manufacture or distribution that may cause a wholly owned or affiliated transporter to favor some goods over others. A transporter can assure a stable revenue stream by selling services in bulk and, over time, to a particular shipper in a way that may prevent a competitor from having periodic access. Or it may favor large shippers over small ones. Policies of this type are not permitted by common carriers as a point of common law.[27]

Granting Transport Routes and Issuing Franchises

The modern practice of granting transport franchises or licenses originated in British common-law efforts to promote inland transportation during the Industrial Revolution in the nineteenth century.[28] The granting of franchises promoted this transport in three ways. The first was the matching of risks and rewards for rendering such service. To the extent that franchises afforded transport companies a certain exclusivity and right to profits for the business, they encouraged inland transport operators to take greater risks in the development of their services. Second, franchises promoted regularly scheduled service to a degree that the marketplace itself might not provide. Third, franchises provided access to land owned by other private interests, using the eminent domain powers of the government.[29] That is, the obligations to accept the risks inherent in the enterprise, keep regular schedules, and accept all comers when space was available accompanied the government's exercise of its ability to take the land needed to promote the service. In other words, the government clears the way for the service as long as the transport firm accepts particular obligations to provide it.

Who gets to build the new pipeline? Is the choice made through a process of political gamesmanship or genuine competitive rivalry? Do incumbents have the ability to erect barriers to entry by drawing on current captive customers, or do both potential entrants and incumbents compete equally to secure the business of new pipeline customers? The respective oil and gas pipeline legislation in the United States eventually answered these questions in different ways, based on whether Congress applied boilerplate common carriage language or not, with great consequences for the structure and performance of their respective industries.

The Tension between Common Carriage and Asset Specificity

These long-standing conceptual foundations for common carriage run into problems with pipelines' asset specificity, as many forms of price or access discrimination among users are not allowed under the traditional legal definition. The nineteenth-century US railroads, with their great need for unheard-of quantities of capital to sink into rights-of-way and rails, faced a number of difficulties in maintaining their profitability without exercising selective forms of price discrimination. In particular, railroads faced highly elastic demand in the long-distance market and inelastic demand for short hauls. The railroad companies also found the need to price-discriminate given the bulky nature of most west-to-east freight shipments and the much higher value, low-bulk east-to-west ship-ments (which otherwise caused empty boxcars to pile up on the East-ern Seaboard).[30] With great differences in demand, the need to promote east-to-west bulk traffic, and the overarching need to try to recapture from shippers and consumers some of the economic rent generated by these new transport routes—as a way to repay investors—the railroads tried not only price discrimination but also various forms of pooling and collusive agreements. It was a turbulent time of rate discrimination, complaints from shippers, and attempts by the states (as opposed to the federal government) to control railroad practices.

The turbulence generally came to an end in the decade after Congress passed the Interstate Commerce Act in 1887 in order to prohibit "un-due and unreasonable preference" and to prevent price discrimination "under substantially similar circumstances and conditions."[31] Although the act was bitterly opposed by railroad managers, those managers came to appreciate that the statute, with its common carriage prohibitions against discrimination, could be used effectively to enforce published rate schedules and eliminate price cutting and railroad rivalry. As Da-vis and North put it, "The Interstate Commerce Commission became an effective government-underwritten cartel device in the railroad indus-try, although it had been invented for quite a different purpose."[32] The blanket prohibition against discrimination made life for the incumbent railroads immeasurably easier. Thereafter, and for most of the twentieth century, the railroads became firm believers in the comforts of regula-tion—until the industry's collapse by the 1970s in the face of intermodal competition.[33]

The problems associated with common carriage regulation of pipe-lines are at least as troublesome as those for railroads. Common carriage in such instances is a problem when one is dealing with great commit-ments of immobile capital that have no other use than to transport oil or gas from one point to another. Contractual hazards already incentivized vertical integration of pipeline transport when the industry was young. Vertical integration, however, is not inevitable. Contractual safeguards can be put into place to attempt to minimize such hazards. Indeed, trans-action cost economics covers a spectrum of possible governance struc-tures. As complexity increases or transparency decreases, transactions move from the market to more complex contractual arrangements—and finally to vertical integration.[34] What common carriage ultimately does is to limit the ability of producers, shippers, and pipelines to mitigate those risks—in effect making vertical integration the only tool by which to address issues related to asset specificity.

The problem is even more critical for gas pipelines, as pipeline users cannot as readily or economically hedge against the unavailability of the pipeline or variable demands by storing the fuel at either end. The effect of imposing common carriage on US oil pipelines, as opposed to the specific rejection of common carriage for US gas pipelines, is a key topic for chapters 6 and 7. Those chapters describe how two largely similar pipeline systems evolved so differently as the result of embracing the institution of common carriage for one and rejecting it for the other.

Evolving Regulatory Institutions

North's recognition among economists will always be identified with his ambitious work on investigating how the search for economic gain spurs institutional change through expanded property rights and reduced transaction costs—and that, indeed, in certain instances, changing institutions can do more to promote economic growth than new technology or factors of production. He argued, with Lance Davis, that changing methods for capital accumulation that accompanied railroad development, along with the poor history of public funding for canals, changed the relative cost of public versus private capital and effectively ended governments' financial participation in major inland transport projects in the United States. They described how, early in the twentieth century, "enlightened" railroad managers essentially hijacked well-intentioned legislation (the Interstate Commerce Act of 1887, the Elkins Act of 1903, and eventually the Hepburn Amendment of 1906) to further consolidate the cartelization of US railroads and prevent discriminatory departures from published rate schedules.[35] This book charts what happened *next* regarding the institutions for major inland transportation projects in the United States.

The controversy among economists about the role of evolving governance institutions is a good deal smaller a decade into the twenty-first century than it was when some of the profession considered such an institutional perspective merely a change in emphasis or an extension of "law and economics."[36] This modern viewpoint remains particularly useful for the pipeline industry, with all its relatively simple and unchanging technology. It is a business tailor made for studying how institutions evolve. It may help, however, to discuss three kinds of institutional development (which appear in the United States and elsewhere) and preview some major events that either stunted or advanced the growth of pipelines as sources of competitive inland transport.

The Courts

From the perspective of a comparative analysis of pipeline development and the economic governance institutions surrounding them, there is no getting around the role of the US Supreme Court. Nobody up to John R. Commons's time had studied the source of the Supreme Court's legal and economic theory in greater depth than he did, and certainly nobody waxed more poetic on that institution's central role in promoting capitalism, when he called it "the first authoritative faculty of political economy in the world's history."[37]

Perhaps the Supreme Court's greatest contribution to the development of US pipelines was its half-century struggle with the meaning of regulatory value that the courts would use, as the numeraire for private property, under the US Constitution. The court first tried to be definitive on the matter in 1898 in the appeal of a regulatory matter called *Smyth v. Ames*.[38] It failed in that task, however, and for the next four decades, investors, commissions, and users of regulated services battled with one another over obscure methods of property valuation for regulated tariffs. The court's failure to resolve the matter came to a head in the legal challenge (by a Standard Oil affiliate) to the authority of the new federal agency (the Federal Power Commission—later the Federal Energy Regulatory Commission) to limit gas pipeline tariffs—called the *Hope Natural Gas* case. With great insight and even greater consequences for US regulation, the court changed the basic focus of regulation from reasonable *property values* to reasonable *earnings*.

A legal system's ability to define and safeguard property rights is fundamental to defining the bundle of legal entitlements that form the basis for Coasian bargaining for pipeline capacity. In the economic literature, there are two meanings for the term *property rights*. The first pertains to the wide meaning of traditional English/American common law and refers to any tangible or intangible rights related to physical ownership, patents, copyrights, and, most important for pipelines, contract rights.[39]

The second is the Roman/civil law definition, which some modern economists writing about the subject interpret as being restricted to physical
objects or tangibles.[40] As is the case with many terms in an institutional
analysis of regulation, the common-law definition does not generalize
easily around the world. Commons noted in 1934 the difficulty that US
and European economists have when dealing with different legal systems, as well as the problems that come when governments' legislative
branches are superior to their judiciary, in ways just as relevant in the
twenty-first century.[41]

For the international community of economists, it is hard to overstate the central relevance of the divergent points of view that Commons
pointed out in the 1930s. The evolution of US pipeline regulatory institutions is fundamentally affected by judicial precedent. In the United
States, regulatory statutes passed by Congress do not truly become law
until interpreted by the Supreme Court in a particular dispute, which
then becomes precedent for further disputes. The discussion of US pipeline regulatory institutions places great emphasis on the finality associated with decisions made and accompanying reasoning given by the Supreme Court, which is superior in such questions to both the legislative
and executive branches of government.[42]

The property rights that create the basis for the competitive market in
legal entitlements are a function of the US common law. Such property
rights may not generalize easily to European civil law jurisdictions.[43]
Contract enforcement is a critical feature of property rights in capacity.
The institutional features of the different major legal systems (German
civil law, French civil law, Scandinavian civil law, and English/American
common law) feature prominently in whether the rights that create the
basis of Coasian bargaining in the United States could be enforced for
pipeline transport capacity in Europe.[44] There seems yet to be no economic consensus on the matter.

The Legislatures

Contending parties frequently enough appealed their cases to the US
Supreme Court, a body that could decisively reflect current politics and
public opinion in its interpretation of the US Constitution without Congress's persistent necessity for compromise. Evidence of Congress's at-
tachment to hard-fought legislative compromises is that its 1906 and

1938 statutes for oil and gas pipelines remain in force. Despite its seeming fondness for existing pipeline legislation—especially after such legislation has been tested in a major appeal before the Supreme Court—Congress has nevertheless shown that it is capable of making new rules for new businesses, learning from its mistakes when the circumstances are right. The fundamentally different foundation for Congress's 1906 and 1938 pipeline laws is the case in point.

The pressures on legislators in other jurisdictions, particularly Europe, appear to be different. For example, the European Parliament has dealt with three "legislative packages" for the European Union's gas pipelines in relatively rapid succession. The federal Parliament in Canberra, Australia, has shown itself capable of making relatively rapid changes to regulatory laws to satisfy idiosyncratic state interests.[45] In both cases, basic new legislation for pipelines seems less destined for court appeal and less permanent in shaping pipeline ownership and/or contractual arrangements.

The Inventors of Novel Institutions

There are times when clever individuals or groups create new governance institutions to solve problems that existing legal or financial customs cannot deal with.[46] Such an invention motivated funds for gas pipelines in the United States after tough legislation in 1935 (the Public Utility Holding Company Act) effectively closed off vertically integrated sources of funding for gas pipelines. During the gas industry's rapid postwar expansion around 1950, the great majority of gas pipeline bonds were held by life insurance companies and other "trustee investments," such as private pension funds, that looked to the life insurance industry for guidance. That funding source did not exist before 1935, as such prospective lenders saw gas pipelines at the time as "unseasoned" investments in a "wasting asset" (the gas itself from the well). Life insurance companies required a satisfactory earnings record over a ten-year period as a prerequisite for investment. The robust new Natural Gas Act prompted fresh study by the investment officers at key insurers.[47] As a result, insurers were willing to accept the new federal regulation as security on long-term loans to the pipeline industry, knowing through careful study that the value of their loans was safely embedded in the uniform accounting and tariff-making system.

The Problem of Timing

If one wishes to create an evolutionary timeline of the institutional elements that create the basis for pipeline regulation around the world, a problem arises. The evolution for oil is nothing like the evolution for gas pipeline regulation, and there is no common evolutionary pattern for North American (US and Canadian) regulation and that in Europe or elsewhere in the world. What timeline might be drawn would show important legislation and court decisions for the half century starting in 1898 and ending in the 1940s. It would then show a gap of four decades followed by a burst of activity toward the end of the twentieth century. Other than the late-century evolutionary burst, the entire remainder of the timeline is all North American and all gas pipelines. Even then, it would be a confused timeline (e.g., the specific governing legislation for US oil pipelines froze after 1906). Outside of North America, the institutional history of regulation for investor-owned pipelines goes back no further than the privatization of British Gas in 1986. The most compelling part of such a timeline would be its empty center. From the mid-1940s through to the mid-1980s, nothing happened. In those four decades—representing two generations of economists—institutional evolution ceased. For modern economists, institutional regulatory memories are shaky. Those who were important in the legislative and court battles that defined the development of US regulation, which are critical to understanding modern US pipeline markets, are long dead and their works are out of print.[48]

The result is that the economists of the late twentieth century on the two sides of the Atlantic (or Pacific) habitually speak past one another, reflecting highly dissimilar economic traditions associated with industry regulation. Those in the United States assumed that the property, administrative, and accounting institutions that they inherited were universal—or at least that the wisdom of their application should be universally obvious. Those elsewhere, less weighed down by legal and administrative history—and having none of Americans' somewhat singular reverence for however their Supreme Court happens to rule at the moment—saw no such universality. Thus, starting with the British Gas privatization, development of regulatory practices began anew and evolved in an entirely new direction—no regulatory accounting, no codified administrative procedures, and vastly more power vested in particular regulators to affect the value of regulated private property without the sort of due process taken for granted in the United States.[49] From North's perspective that economic governance institutions evolve for a reason, the bipolar development of the world's regulatory institutions is a particular complexity that in part motivated this book. Regulatory practice inside and outside North America may converge yet. But with the glacial pace of evolution in regulatory systems so manifestly evident, the basis for economic analysis of regulation should surely shift away from prices and costs (the neoclassical units of analysis) and toward the underlying institutions themselves.

The Theory of Collective Action

The new institutions that modern pipelines illustrate were accompanied by new property rights and new ways of economizing on transaction costs. But the changes did not happen smoothly or organically. Some constituent group pushed, either by capturing the attention of influential legislators (who could propose and pass new legislation) or by prevailing in head-to-head contests before regulators or in the courts.

Those economists with a close connection to the institutions of economic governance have long been dissatisfied with the idea that markets take a smooth path toward equilibrium. They looked at *how* markets move from one state to another. Commons was not convinced that transacting would be harmonious. He thought that order would emerge from conflict among parties, which would ultimately lead to the US Supreme Court.[50] The conflict that defined both the shaping of pipeline legislation in Congress and the legal disputes afterward reflects the kind of conflict/resolution that Commons found in other markets in his studies early in the twentieth century.

Mancur Olson also studied the contests among parties with differing objectives. To Olson, buyers and sellers in markets gained advantages over each other by creating pressure for public policy in their favor. The effectiveness of these groups depends on their sizes and compositions—small and tightly organized groups tend to be effective compared to large and dispersed groups. Contrary to traditional economic theory, neither Commons nor Olson believed that individuals with common interests tended automatically to act to further those interests. Commons saw pressure groups as critical components of markets, particularly when those markets required legislation. Olson investigated what characteristics of pressure groups allowed them to be effective levers for favorable public policy.[51]

Commons used the term *collective action* extensively to describe how organizations work collectively to control the actions of individuals or corporations. Commons defined an economic institution as "collective action in control of individual action."[52] Collective action has been described in a roughly equivalent fashion by modern institutional economists.[53] The actions of the pressure groups involved with pipelines—owners, petroleum producers, shippers—shape the institutions and industrial organization of pipelines in a way that the neoclassical theories of production cost do not analyze. Commons thought that in each transaction, there was the potential for conflict, which only a mutual dependence of the parties and a desire for order, through collective action, would overcome. He placed the Supreme Court in a unique position when it came to dealing with the work of pressure groups. To him, what the court ultimately wants "is not truth but orderly action. The concern must keep agoing."[54] The pursuit of orderly action in the collective control of individual firms was a major theme of Commons's work, as recognized much later by Williamson.[55]

Commons tried, and ultimately failed, to create an economic theory in which his view of collective action played an essential part. As Geoffrey Hodgson, from the University of Hertfordshire, says, "In America in the 1930s, the influence and momentum of [Commons's] institutionalism were themselves sufficient to ensure its prominence and survival for at least two more decades."[56] But except for a few institutional economists and those at the University of Wisconsin, like Martin Glaeser, and their own students, Commons's role in trying to create a theory of collective action was largely forgotten.[57]

Mancur Olson came three decades later. The base of his analysis was that small consumers in large markets would have a problem expending any individual effort to pursue an outcome that would benefit the whole group. That is, in such situations, an individual's own effort in a group had the character of a "public good," which no individual would rationally expend effort to achieve. Olson proved this proposition using the mathematics of basic and uncontroversial game theory, taking this result as evidence that economists had long overlooked something that should have been obvious.[58] Further, he saw the practical application of his proposition appear repeatedly in the relations between producers and consumers in markets. While consumers would have an interest in opposition legislation or regulations that would produce supracompetitive prices for producers, and would have an interest in creating buyers' co-alitions to prevent such policies, he found no major country where most consumers are part of any organization that works to their common in-terest. Looking back to the study of pressure groups in and prior to the early twentieth century, Olson saw that the study of collective action went back to the beginnings of economics but then came to be "strangely neglected during most of the rest of the history of the subject." [59] He at-tributed this neglect to the emphasis in recent times by economists in-terested in the logic for the case of competitive markets, where in reality pressure groups play a role in the logic of market failure when regulators and legislators become involved. Since Olson's initial contributions, text-books and survey papers have been devoted to political choice models that investigate the effect of collective action on public policy.[60]

In generalizing his logic of collective action—that the larger the num-ber of individuals in a group, the less able it is to act in the group's com-mon interest—Olson drew a number of implications that appear to be important in how groups press to shape pipeline regulation and mar-kets. For such interest groups as pipelines or fuel producers, which re-flect a narrow segment of society, Olson holds that there is no real con-straint on the social cost that they will find it expedient to impose on society in the course of obtaining a larger share of the social output for themselves—what Olson refers to as "rent-seeking."[61] When the vari-ous actions of those pressure groups comprising pipelines or producers, and the opposing actions of coalitions of gas distributors in the United States for consumers during the 1950s through the 1990s, are witnessed, Olson's implications from the logic of collective action seem to be con-firmed often enough.

Property Rights and Coasian Bargaining

Transaction cost economics would have been vital to explaining pipe-line development, even before 1960. But Coase's introduction of the pos-sibilities for a market in well-defined legal entitlements, rather than the market in merely tangible commodities or services, created another de-velopment possibility—to free point-to-point pipeline *transport capacity markets* from *pipeline ownership*, greatly reducing the regulatory bur-den (other than to enforce the property rights and fully informed and frictionless markets). The shift from regulating pipelines to regulating the rights to legal transport entitlements is a phenomenon explained by transaction cost economics. It lies beyond the frame of reference of traditional neoclassical economics and its focus on the structure of production cost.

The notion of property rights to point-to-point pipeline transport capacity is central to understanding modern pipeline markets. Shippers in the United States have the right to use or sell these rights at unregulated prices on organized exchanges, either themselves or (since 2008) through any agent they hire to do so. Further, the cost of these rights is a well-known function of regulatory rate-making procedures that tie regulated pipeline rates to the specific facilities used to support the capacity rights. Creating these property rights in pipeline capacity in the United States was a contentious task, involving much regulatory litigation. The discussion in chapter 7 shows how the cumulative victories by groups of gas distributors—in litigation first against gas producers and later against the pipeline companies—led to the series of conditions that made possible the defining, safeguarding, and trading of such property rights.

CHAPTER SIX

Transacting with Common Carriage: The Oil Pipeline Regulations of 1906

In 1906, Congress imposed common carriage regulation on the oil pipeline industry (as it had done for the railroad industry in 1887) and put the Interstate Commerce Commission (ICC) in charge of regulating pipelines. The motivation for congressional action was a broad-based and popular assault, led by President Theodore Roosevelt personally, on the way that the Standard Oil Company used both rail and pipeline transport to solidify its dominance of the nation's petroleum industry. The 1906 legislation was highly noteworthy: it gave a federal body the power to act on its own authority to remedy unjust and unreasonable practices on the part of interstate pipeline companies, limited only by the US Constitution's general safeguards regarding private property and due process.

From the perspective of orderly and efficient pipeline regulation, however, the 1906 legislation was a failure. First, the mechanism traditionally chosen to regulate transport businesses in the nineteenth century, which that legislation reflected, precluded the development of an independent pipeline industry by prohibiting contractual commitments between pipelines and shippers. Deprived of the ability to limit access by contract, pipelines presented too great an investment risk for the producers and/ or refiners (who spurred their construction) if they could not be assured of privileged access to those lines through vertical integration coupled with various methods of discriminating effectively against independent shippers. Second, the ICC in 1906 had neither the tools to regulate pipeline prices and access terms effectively—such regulatory tools had not yet been invented—nor the obligation to try to do so anyway. As a result, from 1906 until the agency's dissolution by Congress in 1978, the ICC initiated no action of substance with respect to the orderly and effective regulation of oil pipelines.[1] The industry's post-1978 regulator, the Federal Energy Regulatory Commission (FERC), is familiar with those regulatory tools and has created some order in oil pipeline tariff setting. But the industry is still bound by that century-old common carriage legislation and remains dominated by vertically integrated, joint-ventured pipelines.

The problems with the initial pipeline legislation illustrate why the common carriage instrument chosen by Congress—however useful it may have been for rail or road transport—had no chance of succeeding in an industry whose players committed great quantities of capital at the production, transport, and refining stages with such asset specificity. The debates in Congress reveal the frustration that individual senators or representatives felt in applying the regulatory tools then at hand. These debates, which occurred more than a century ago, showed instances of deep insight and wise practical judgment of some of those on the floor of Congress. While those who debated the matter did not have modern economists to advise them, it is clear that they were nevertheless guided by the economic principles that shape the way that pipelines transact.

Common Carriage Imposed on Standard Oil's Pipelines

By the early twentieth century, practically all major American oil pipelines had been absorbed by the Standard Oil Company. The writing of contracts during that period was undoubtedly difficult, given the complications involved in obtaining reliable and verified business information about companies, shipments, and costs. But it would be incorrect to conclude that communications problems were the driving force behind the integration of pipelines into Standard Oil. The reason for Standard Oil's absorption of both pipelines and railroads was the company's desire to cement its dominance in the petroleum market and to bar entry to competitors.[2] The Garfield report, which compelled Congress to regulate oil pipelines at the federal level, demonstrated as much.[3]

On May 4, 1906, the US Senate took up the question of bringing pipelines under the jurisdiction of the ICC. President Roosevelt opened the Senate debate by transmitting an unusual letter to the Senate floor. Roosevelt's letter cataloged the Standard Oil Company's exploitative rail transportation practices and underlined his strong recommendation for a legislative remedy given the "numerous evils which are inevitable under a system in which the big shipper and the railroad are left free to crush out all individual initiative and all power of independent action because of the absence of adequate and thorough-going governmental control."[4] With respect to the preexisting ability of aggrieved shippers to obtain legal, as opposed to regulatory, redress for such problems, Roosevelt held that the "instrumentality of a lawsuit" was an inadequate remedy. What he recommended was a commission of "ample affirmative power, so conferred as to make its decisions take effect at once, subject only to such action by the court as is demanded by the Constitution."[5]

Roosevelt's letter and the public's ire were specifically directed toward the railroads; his letter did not refer to pipelines. In fact, the Hepburn bill, which passed in the House of Representatives three months before the Senate debate, had no provision for pipelines.[6] Immediately after Roosevelt's letter was read on the Senate floor, Republican senator Henry Cabot Lodge of Massachusetts introduced a well-orchestrated amendment to the bill to address the oversight. Lodge's proposed amendment brought oil pipelines also under the jurisdiction of the ICC. In extending the Hepburn Amendment to oil pipelines, the Lodge Amendment indicated the exclusion of gas pipelines, a possible sign that Lodge was trying to maneuver around potential opposition to the bill. The ensuing Senate debate—on whether to exempt gas pipelines from the proposed legislation—was to separate oil from gas pipeline regulation for the next century and beyond. The most vocal proponent for that exemption was Senator Joseph P. Foraker (R-Ohio), from Roosevelt's own party. His single-handed and indefatigable opposition to including gas pipelines in the Hepburn Amendment charted a very different course for the future regulation of gas pipelines.

Gas Pipelines Escape Common Carriage

No senator questioned the wisdom of bringing oil pipelines under ICC jurisdiction. Gas pipelines were a different matter, as they were newer, smaller, and not so central to Standard Oil's operations. Senator Ben

1908

FIGURE 4. Two views of Senator Joseph P. Foraker of Ohio, the ironic champion for modern competitive pipeline transport markets (Above is the senator's picture from his 1916 memoir. On the facing page is a cartoon from the cover of *Harper's Weekly* from November 25, 1905 (when the Hepburn Amendment was being debated). President Roosevelt is seen spurring on the Senate Republicans against a locomotive masquerading as an elephant. A tireless railroad advocate, Foraker was the only Republican ultimately to vote against the full Hepburn Amendment. *Harper's* readers would have known that Foraker is the object of Roosevelt's attack.)

HARPER'S WEEKLY
JOURNAL OF CIVILIZATION

VOL. XLIX   *New York, Saturday, November 25, 1905*   NO. 2551

TO A FINISH

("Pitchfork Ben") Tillman (D–South Carolina) contended that the essential test related to interstate commerce itself: if a pipeline carried goods between states, whether it was oil or gas, it should be regulated by the ICC. Foraker claimed that gas pipelines by their nature were not common carriers and that, particularly for gas pipelines, binding gas lines with common carrier obligations would imperil funding for a useful and profitable private business.[7]

Senator Porter McCumber (R–North Dakota), generally supporting Foraker, made a distinction between oil and gas pipelines based on the close relationship between gas pipelines and the gas they carried. He drew attention to the dedicated, private nature of gas pipeline shipments, saying, "If [gas pipelines] are carrying their own goods and no goods of the public, how is the public interested one way or the other in the matter of their carrying their own goods through their own pipe line from one State to another?"[8] Tillman responded with a concern about how a pipeline company could purchase gas from one landowner but simultaneously draw upon the pressure from surrounding landowners not connected to the pipeline.[9] Foraker responded to Tillman that it was the capital involved that was the key issue, for common carriage would give any landowner the ability to access the pipeline without committing to the capital expenditures needed to build it.[10] Foraker argued that attempts to secure financing for the gas line would fail if the line were considered a carrier of gas rather than a means to secure a distant state's gas for private purposes.[11] Tillman tried once again to voice his notion of the injustice of failing to allow other gas producers to secure access to a line as a common carrier, saying that the "first possessor of gas will have a monopoly and will extract tribute from every man who has gas under this land and who bores a well."[12] Foraker kept at it, saying that "there is no other customer there except only this one company, which will own the pipe line. It is not our purpose to engage in the business of common carriage, and I do not think we ought to be treated as a common carrier when we are not."[13]

And that was that. Foraker outlasted Tillman, and he ended forever the debate over whether to classify US gas pipelines as common carriers. It was a critical crossroads for the future of the US gas industry. It seemed clear to the Senate that the pipeline Foraker described was for the private use of the gas company in Cincinnati. The Senate was also convinced that the concern about other landowners having their gas pressure drawn away by the first driller in a field was a matter for the states to deal with, not the federal government. Mostly, the Senate was evidently convinced that those financing the line needed to know that it was theirs to use. After almost fourteen pages of courtly debate, the Senate voted unanimously to exempt gas pipelines from the Hepburn bill. The debate dealt squarely with the practical nature of pipeline transacting and asset specificity in a world where the only known regulatory tools for dealing with transporters were those of common carriage. It may have been clear to those present that imposing common carriage made little more sense for oil pipelines than for gas pipelines. But no senator in 1906, just after Ida Tarbell's massive and popular exposé of John D. Rockefeller, would have been willing to stick his neck out for the sake of Standard Oil, which controlled those oil pipelines. Such was the point in time that locked in everything for oil pipelines for the next century. History matters indeed.

When Congress eventually turned its attention to federal regulations for gas pipelines in 1938, it did so with a vastly more sophisticated set of regulatory tools. It gave federal regulators licensing authority and provided for accounting, rate-making, and administrative methods learned in the meantime at the state regulatory level—tested by the courts and widely accepted. It is no overstatement to call Senator Foraker the father of the modern competitive gas pipeline transport markets. He might have been amused.[14]

The "Commodities Clause" and Vertical Integration

The next critical topic at hand in the evolution of pipeline regulation, raised three days after the Senate debate over the Lodge Amendment, concerned the "commodities clause" forbidding transport companies from owning the commodities they carried. The clause was an attempt by Congress to break the power of the railroad companies in Appalachian coal fields. Senator Stephen Elkins, a Republican from West Virginia coal country, brought the commodities clause amendment to the Senate floor, saying, "I want to confine the railroads to the legitimate business for which they are incorporated—the transportation of freight and passengers."[15]

With considerable foresight that would impress even twenty-first-century economists, Senator Knute Nelson (R-Minnesota) wished to impose the commodities clause on pipelines as well. He foresaw the futility in trying to regulate a pipeline's charges when the pipeline also owned the fuel it shipped.[16] Nelson's point was central to how both gas and oil pipelines would be regulated thereafter. If the commodities clause were not applied to oil pipelines, then pipelines would be closely integrated (as they were at that time) into production and refining. Pipeline rates would simply be a part of final delivered and marketed oil products—and thus merely internal company "paper" transactions. But Senator Chester Long (R-Kansas) saw practical problems in light of Standard Oil Company's as yet unbroken dominance in refining and marketing. He argued that Standard Oil could easily work around the commodities clause by permitting affiliate operations to take possession of the oil in other affiliates' pipelines. So while Long supported subjecting pipelines to the commodities clause in theory, for the reasons voiced by Nelson, he vehemently opposed such a move in practice. He warned his colleagues that imposing the commodities clause in that environment would wreck the hard-pressed independent oil business, and he convinced his colleagues not to apply it.[17]

From 1906 on, oil pipelines—these relationship-specific investments—were regulated as common carriers that were obligated to provide transport service to all comers. In practice, this meant that most oil pipeline ventures would develop as vertically integrated companies and later combine into joint ventures among the major oil companies. Senator Long was correct: oil pipelines were not conceived of, or built as, common carriers in the traditional sense. They dealt with the risk posed by asset specificity by affiliating with the producers and refiners that would use the line. But Senator Nelson was also correct: regulating pipeline prices without the commodities clause would not be a practical way to control objectionable behavior in the petroleum markets.

US Petroleum Industry Adjusts to Common Carriage without the Commodities Clause[18]

The result of the 1906 Hepburn Amendment, coupled with the antitrust breakup of Standard Oil in 1911, was the rapid creation of a number of independent oil pipeline companies. That independence did not last. By the late 1930s, all but one interstate oil pipeline had either failed or once again become part of a vertically integrated petroleum company. That evolution back toward vertical integration prompted the Department of

Justice (DOJ) to bring a major antitrust case that was only set aside with a quick settlement in December 1941 in light of the US entry into World War II. That settlement took the industry up to 1978, when Congress dissolved the ICC.

The ICC Fights Standard Oil to Assume Jurisdiction: 1906–14

Immediately after the Hepburn Amendment's passage in 1906, the ICC, swamped by its rail-related workload, devoted little time to oil pipelines. The ICC's first set of uniform accounting regulations, created in 1908, applied only to railroads. Not until 1911 did the ICC turn its attention to accounting reporting requirements for the oil pipelines. For its part, the Standard Oil Company, which controlled most of the oil pipelines, was unsure how to deal with the ICC's newly enhanced authority. Accepting and complying with the Hepburn Amendment without a fight was unthinkable. For Standard Oil, the amendment violated long-standing beliefs about the sanctity of private property. The company therefore pursed a two-part strategy: attempt to evade ICC jurisdiction (by organizing its pipeline operations so as technically to avoid the letter of interstate commerce law) and attack the Hepburn Amendment's legality in the courts.

The Hepburn Amendment was vague on whether the ICC could initiate investigations to determine whether existing oil pipeline rates were unjust, unreasonable, or unduly discriminatory. Another act of Congress was required to give the ICC specific authority to initiate rate proceedings.[19] After the ICC issued its order for pipelines to file tariffs in 1912, lawyers for Standard Oil appealed to the US Commerce Court, which disagreed with the ICC's ruling.[20] Presiding justice Martin Knapp (formerly of the ICC) delivered the opinion that "it seems to us too plain for argument that these private pipe lines cannot be legislated into public facilities, and that the [Hepburn] amendment necessarily deprives the owners of such lines of their property rights without just compensation."[21] The ICC promptly appealed to the Supreme Court, where, in a judgment delivered by Justice Oliver Wendell Holmes in 1914, the Supreme Court reversed Knapp's decision and ended any ambiguity about the ICC's regulatory authority over oil pipelines.[22]

Eight years of purposeful delay and legal process passed between Congress's passage of the Hepburn Amendment and the confirmation by the Supreme Court that the ICC did in fact have jurisdiction over oil pipeline rates. For the ICC, obtaining jurisdiction was one thing—using it was quite another. Standard Oil was subsequently dismembered by antitrust action, the oil pipeline industry structure did change (and then change back), and ultimately pipeline owners and the US government agreed on a significant settlement regarding rates. But the ICC took no active role in promoting any of these developments.

The Newly Independent Pipelines Eventually Reintegrate: 1911–31

Over the course of twenty years, the oil pipeline industry split apart from oil production—for reasons of corporate control and a wider worry about government intrusion into business affairs—only to reintegrate as the only practical response to a common carriage statute that removed other options for dealing with asset specificity. Between 1911 and 1914, the oil pipeline industry rapidly restructured itself into a group of separate and independent businesses. The industry-wide disaggregation began when, on May 15, 1911, the US Supreme Court issued its first major decision under the Sherman Antitrust Act and ordered the breakup of the Standard Oil Company. The Supreme Court's unanimous decision mandated the corporate separation of ten common carrier pipelines and three partially or wholly integrated oil companies that owned pipelines. After the Supreme Court's 1914 decision solidifying the ICC's jurisdiction, Standard Oil, which had remained in control of several newly separated pipelines after the 1911 Supreme Court–ordered breakup, divested its crude oil pipelines into newly incorporated units. Standard Oil's move was prompted by concern that the 1914 decision foretold further government intervention into its production and refining businesses.

The swift corporate divestiture of the common carrier pipelines had no initial effect on the new pipeline companies' functional relationship with their connected oil producers and refiners. Between 1914 and 1931, however, a definitive trend led to the decline of the newly independent oil pipeline companies as they reintegrated with producers. Even Prairie Pipe Line Company, the largest pipeline operating in the United States at the time, failed to survive as an independent pipeline company. Only one major oil pipeline, the Buckeye Pipe Line Company, survived the period as an independent transporter.

Prairie Oil and Gas Company had been the Standard Oil Company's affiliate for purchasing, producing, and transporting oil in and from the Mid-continent Oil Field of Oklahoma and Kansas. At the time of its separation from Standard Oil in 1911, Prairie Oil and Gas was the largest buyer of crude oil in the Mid-continent region for transport to refineries in Kansas City and Chicago. In 1914, the company owned over five thousand miles of trunk and gathering lines and was the largest pipeline system in the country, shipping thirty-nine million barrels of crude oil that year. Prairie Pipe Line Company's management pursued an aggressive expansion program during World War I and immediately after. Throughout this period, Prairie Pipe Line enjoyed the advantage of being the sole pipeline running between the Mid-continent Oil Field to the northern and eastern regions of the United States.[23]

Throughout the 1920s, Prairie Pipe Line tried to retain its shipments of Mid-continent oil despite the defection of many customers—some to tanker traffic through the Gulf of Mexico to the East Coast. The loss of major customers like Standard (New Jersey) hurt Prairie's connecting pipeline carriers to the East Coast. With shipments declining as refinery customers looked elsewhere for crude oil, Prairie Pipe Line and its affiliate Prairie Oil and Gas were absorbed by Sinclair Consolidated and placed into a new entity called Consolidated Oil Corporation, an integrated petroleum company. Ultimately, the basic problem was that there was no contractual ability under common carriage for Prairie to resist the pull of formal vertical integration to assure the utilization of its pipeline assets in an area of increasing pipeline rivalry and intermodal tanker competition.

Buckeye Pipe Line Company faced a different fate. Buckeye anchored the set of Standard Oil pipelines known as the Northern Group, whose lines stretched east from Chicago to New York State. Buckeye separated from Standard Oil in the 1911 dissolution decree. It occupied a strategic geographic and market position in two respects: its role as a transshipper of Mid-continent oil to the East Coast and its handling of Ohio oil production and service to Ohio's refineries. Both roles enabled Buckeye to remain viable despite tanker competition from the East Coast refineries. Integrated oil companies increased their refinery operations in Ohio during the 1920s, and while they abandoned Prairie in its various territories, they remained customers of Buckeye.

Buckeye, however, is an exception to the era of consolidation. It is the only major oil pipeline company that survived as an independent entity throughout the 1930s and 1940s and up to the present day. To survive, Buckeye had to perform a role that it would not pay large oil firms to duplicate in the absence of the kind of contractual tie forbidden by com-

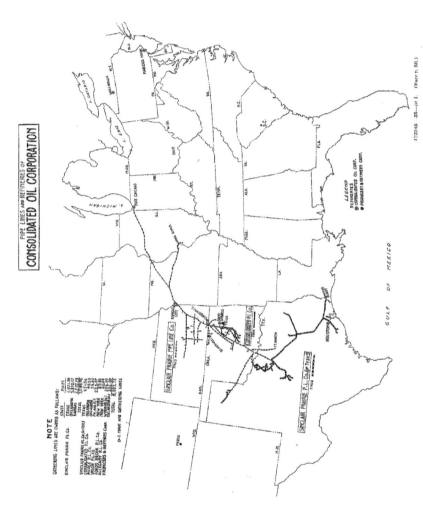

FIGURE 5. Consolidated Oil Corporation absorbs Prairie Pipe Line by 1931 (Reprinted from Splawn, *Report on Pipe Lines*, facing I:50)

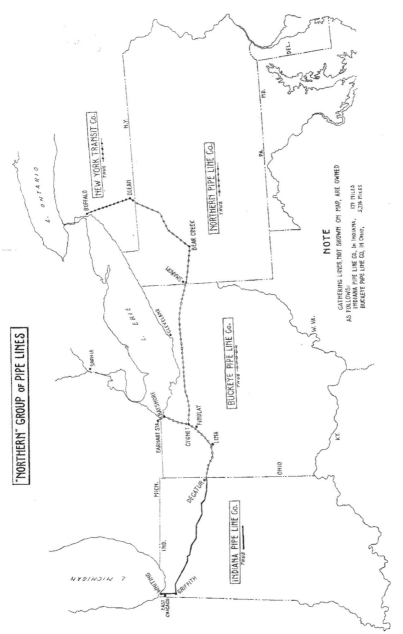

FIGURE 6. Buckeye Pipeline in Ohio alone remains independent (Reprinted from Splawn, *Report on Pipe Lines*, facing 1:389)

mon carriage. Even then, it is unreasonable to think that Buckeye could exist and/or set its charges independently of the wishes of the integrated oil companies it served. For every other independent oil pipeline company, vertical integration was the only way to stay in business.

Public Pressure Leads to the "Consent Decree": 1931–41

By 1931, the industry was vertically integrated once again, and independent oil producers and refiners were complaining to Congress about lack of access to the pipelines owned by integrated oil companies. They called for the "divorcement" of pipelines with every new congressional session.[24] But neither the problem nor their persistent lobbying was sufficient to motivate Congress to create new legislation over the strenuous objections of the integrated oil companies. There were, after all, seemingly evident economies in vertically integrated pipeline operations, and the common carrier nature of pipeline regulation created risks for recouping pipeline costs that large-scale integrated operations appeared to allay.

As the Great Depression worsened, the oil industry found itself with an unexpected oil surplus. As a result, various pipelines instituted new minimum shipment and other exclusionary actions to deny service to small producers. At the same time, the nominal profitability of the integrated pipeline companies soared, with a number of the pipelines companies showing dividend payouts to their oil company parents exceeding the amount they had invested during the period of rapid pipeline growth in the 1920s.[25]

Representative Sam Rayburn of Texas, chair of the Committee on Interstate and Foreign Commerce and future Speaker of the House, seized on the plight of the independent producers. Rayburn asked economist Walter Splawn, of the University of Texas–Austin, to conduct a detailed investigation into the structure and conduct of petroleum pipelines. Splawn's one thousand–page report, completed in 1933, dealt with the corporate ownership, operations, rates, and profitability of US oil pipelines.[26] Splawn looked at the returns earned by pipeline companies, showing that the average return on investment during the period 1921–31 was 14.49 percent, and the return on net investment during the period 1923–31 was 25.43 percent—profits that Splawn considered excessive.

The Splawn report shed light on the oil pipeline business in a way that the Garfield report had done for railroads before the drafting of the Hepburn Amendment. Both reports pointed to the difficulty that independent producers had in obtaining access to vertically integrated common carrier lines. For pipelines, this lack of access gave the integrated refiners the exclusive ability to locate their refining facilities near the markets, leaving the independents to less advantageously located refiners in the production areas.

Ultimately, Splawn's report led to action by the DOJ to bring a Sherman Act case against the integrated oil companies. On September 30, 1940, the DOJ filed suit against twenty-two major oil companies and the industry's trade association, the American Petroleum Institute (API), alleging a conspiracy to fix the price of petroleum transportation.[27] The complaint called for the dissolution of the API and an injunction against the methods used by the oil companies to fix prices, including minimum tenders, onerous and oppressive rates, and other devices that created barriers to the use of vertically integrated pipelines by independent shippers. In addition to the main antitrust action, the Justice Department filed three cases against fifty-nine integrated oil pipelines, alleging violation of the Elkins Act of 1903 prohibiting rebates to affiliates (a misdemeanor violation aimed at the secret Standard Oil rail rebates, with treble damages that could have crippled the industry). The cases were filed under the direction of Assistant Attorney General Thurman Arnold, a former Yale University law professor and a great antitrust champion of the 1930s.[28]

In late 1940 and early 1941, Arnold and the oil companies worked to negotiate a resolution of the two cases. The Elkins cases were settled based on a stipulation that oil pipeline dividends to affiliate owners would be limited to 7 percent of ICC pipeline valuations. By August 1941, however, it was clear to both the DOJ and the industry that the war in Europe and defense preparedness might make it impossible to settle the main case. The attack on Pearl Harbor changed the nature of the negotiations completely. While various parties had already claimed that a December 5 settlement was unsatisfactory, those claims were dropped and the parties returned to the draft that most of the defendants had signed the previous September 16. Without further discussion, Arnold proposed to accept that draft, and the industry agreed unanimously. The provisions of the consent decree prevented common carrier oil pipelines from paying to pipeline owners any more than the 7 percent cap, directly or indirectly, based on ICC valuations.[29]

A Postwar Call for Pipeline Transport Contracts

The end of World War II renewed the calls for divorcement of pipeline transport from oil production and refining. A 1948 book by Eugene Rostow, another Yale law professor, reintroduced the debate over vertical integration in the oil industry to the public. Rostow wrote that the "chief weapon of the major companies for protecting their position in the market for crude oil is their ownership of pipe lines."[30] According to Rostow, rate regulation was an inadequate solution to independent companies' dominance of the sector. The oil pipelines existed, he said, not to make money as common carriers but to ship their builder's oil, and they would give preference to shipping their own oil versus the oil of independent producers, whatever rates the ICC set. In particular, Rostow highlighted the ICC regulatory ineffectiveness without the commodities clause, writing that "Congress has never extended the principle of the Commodities Clause of the Hepburn Act beyond railroads, so that the independent forces in the industry, confronting major company control of the pipe lines, were left to their dubious remedies under the Interstate Commerce Act."[31]

According to Rostow, only by applying the commodities clause to oil pipelines could the petroleum industry be truly competitive. He called for contractual, rather than corporate, links between producers, pipelines, and refiners. He proposed the study of pipeline transport contracts—the first scholar to do so. Considering that elements of the Interstate Commerce Act outlawed discrimination among shippers, Rostow (later with coauthor Arthur Sachs) framed the issues that the ICC and courts would have to face in order to facilitate orderly transacting—by contract—with independent oil pipelines.[32]

Three years later, George S. Wolbert, a young law professor at Washington and Lee University who had previously worked at the Phillips Petroleum Company, published a book rebutting Rostow.[33] In *American Pipe Lines*, Wolbert held that oil pipelines were not built as a means for controlling the crude oil market but rather as a reflection of normal business reactions to the problem of raising money for pipelines, particularly the great "hazard or risk" that the capital would fail to be amortized and the interdependence of pipelines and refineries.[34] Shortly thereafter, a third book appeared entitled *Crude Oil Pipe Lines and Competition in the Oil Industry*, by Leslie Cookenboo, of Rice University.[35] Neither Cookenboo nor Wolbert saw any practical value in the imposition of the commodities clause requiring transport to be separated for oil ownership. Cookenboo's opinions, however, were driven by a preoccupation with economies of scale—to be achieved through compulsory joint ventures if necessary. Neither Wolbert nor Cookenboo addressed whether some form of contracting priority could avoid the risk of transacting inherent in common carriage.

With the exception of Rostow, the writers during this period all took as a given the traditional common carriage prohibition against preferential access to transport facilities, an approach that blocked the pipeline companies from using contracts to discipline access to the lines. Without the ability to form such contracts, the only way to deal with the transaction-specific risks of dedicated pipeline assets for writers like Wolbert and Cookenboo was through vertical integration and joint venturing. In questioning how to structure contract-based transport under the Interstate Commerce Act, Rostow was decades ahead of his time. Other contemporary writers, including Johnson, Wolbert, and Cookenboo, considered Rostow's proposals to be merely academic.[36] Rostow's visionary contract carriage discussion was certainly obscure, buried as it was in two largely theoretical legal tracts. No group was sufficiently organized and funded in the 1950s to clarify or press the ICC to exploit the concept of contract-based transport for the benefit of independent oil pipeline shippers. It took another thirty years for a regulatory application of contract-based transport to come before the FERC, in an initiative that was organized and funded by pressure groups of gas distributors that were willing to pursue Rostow's visionary proposal for gas pipelines.

The Oil Pipeline Business Evolves around the Consent Decree

The consent decree capped dividends paid by oil pipelines at 7 percent of ICC valuation. The immediate result of the consent decree is evident in Cookenboo's chart of oil pipeline returns before and after the decree, reproduced as figure 7. By the late 1940s, after-tax pipeline returns were driven toward 7 percent on ICC valuations, and the average pipeline rate declined. With lower returns, the integrated oil companies experienced a longer payback period for new pipeline investments, increasing the risk that new, postwar oil pipeline investments would not pay.

The lower rates and longer payback periods prompted two moves by the pipelines. The first was to boost pipeline profitability in spite of the cap. Raising effective profitability was relatively straightforward, as the

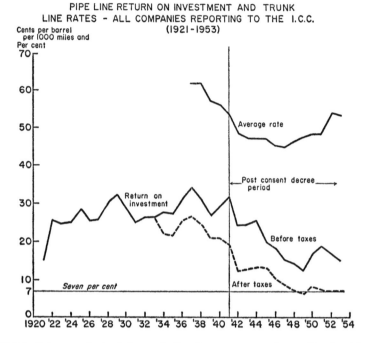

PIPE LINE RETURN ON INVESTMENT AND TRUNK
LINE RATES - ALL COMPANIES REPORTING TO THE I.C.C.
(1921-1953)

FIGURE 7. Returns and rates before and after the 1941 consent decree (Reprinted from Cookenboo, *Crude Oil Pipe Lines*, 99)

7 percent cap was based on total ICC valuations rather than the amount of equity invested. To evade the cap, pipeline companies had merely to display capital structures with a high debt percentage—higher than those displayed by any stand-alone utility. To achieve such leverage levels, integrated oil companies simply guaranteed the debt of pipeline affiliates.[37]

In addition to their main purpose of providing for a backstop payee for debt, the affiliate guarantee also served two other purposes. It ensured that the guarantor exercised a supervisory function over a borrower, and, in certain circumstances, it secured noninterference from the guarantor.[38] The noninterference role is important for joint venture oil pipelines, as it helps to assure that one among the otherwise competitive oil company partners will not work to interfere with the pipeline's operations if it discovers more conveniently placed oil elsewhere. With these guarantees, pipeline operations could borrow heavily and raise the profitability of pipeline ventures for their owners well above the 7 percent cap.[39]

The second move to deal with longer payback period and lower rates was to ensure that shippers would consistently use particular lines in spite of the equal access implied by common carriage. One method involved "throughput and deficiency agreements," whereby each shipper-owner agreed to ship a pro rata share of oil sufficient to allow the pipeline to generate enough cash revenue to service the interest and principal repayment of the pipeline's debt. The obligation was tied to the repayment of debt and lasted as long as the debt was outstanding.[40] Under such agreements, pipelines were guaranteed revenues to service debt, but they had no obligation to reserve capacity strictly for the firms guaranteeing throughput.

Oil pipeline owners also bound shippers by giving priority to those that could show a consistent pattern of yearly shipments. Under these rules, which still exist in the industry, shippers develop a "history" that determines the extent to which they can obtain pipeline space during constrained periods when shipper requests exceed pipeline capacity.[41] The last way to bind shippers to the pipeline was to form joint ventures for the construction of both crude oil and oil products pipelines. Most pipeline projects greater than five hundred miles built between 1946 and 1958 were joint ventures among several major integrated oil companies.[42] Joint ventures continue to be a popular way to create predictable shipments among integrated oil pipelines.[43]

Regulatory Remedies at the FERC: 1985 to the Present

By the mid-1980s, the FERC had not yet addressed the problem of making tariffs for the oil pipelines, which it had inherited from the ICC in 1978. The oil pipelines had no tariff base that would satisfy the FERC's traditional requirements for pipeline rate making.[44] In addition, more than three-fourths of the industry's pipeline capacity was owned by the eighteen major integrated oil companies. There remained substantial overlap in the ownership of both crude oil and oil products pipelines. About 60 percent of pipeline shipments were on lines owned jointly by groups of oil companies comprising their major shippers (which were primarily the major integrated oil companies), which meant that for those lines, the FERC had no ability to hear from an organized, well-funded adverse party that would press the issue of reasonable oil pipeline rate making.[45] The DOJ commented on this intersection of shipper ownership, particularly with respect to the way it co-opted the normal "clash between those regulated and their immediate customers [that] provides the necessary tension to achieve effective and even-handed regulatory scrutiny [and hence] the absence of adverseness requires the regulator to take affirmative steps to regulate effectively."[46]

Given these factors, the FERC faced a difficult regulatory task. If it was to regulate effectively at all, it had to settle the question of the tariff base.[47] At the same time, the FERC had to consider whether crude or products pipelines required continued regulation and whether there were any nontraditional regulatory methods that could help it manage the situation. Prodded by Congress, the FERC began investigating these issues during the 1980s and 1990s.[48]

US Oil Pipelines and the "Absence of Adverseness"

Over the century since the Hepburn Amendment, the oil pipeline industry has exhibited an almost complete absence of effective shipper pressure groups. The regulatory and legislative conflicts arising around oil pipelines were mainly battles between the vertically integrated companies owning oil pipelines and the US Justice Department acting in its role to protect markets from monopoly abuse. Except for the relatively short period immediately after 1914, oil pipelines remained vertically integrated into production and refining. Independent oil producers were widely dispersed price takers in this common carriage arrangement. Both before and after the 1941 consent decree, the integrated pipelines and their trade group, the API, constituted a pipeline- and producer-oriented pressure group with no effective counterpart among independent shippers. The ICC for its part proved to be an indifferent regulatory agency. It was slow to seize jurisdiction, generally uninterested in pipelines (as opposed to railroads), and willing to draw upon API for advice when it found itself in a quandary about how to establish pipeline values for the purpose of setting rates.

Only one major private complaint was leveled against pipeline companies by shippers in the era before the Justice Department filed its lawsuits that led to the 1941 consent decree. The "Bundred decision" involved a formal case brought before the ICC involving minimum tenders. The Bundreds (two brothers) were Pennsylvania-based oil brokers who sought to use the Prairie Pipe Line. They believed that Prairie's 100,000-barrel minimum batch size unduly discriminated against the small shippers and filed a formal compliant with the ICC in 1920. Prairie privately believed that there was no practical, operational foundation for the 100,000-barrel minimum. Nevertheless, fearing that the imposition of any change in its tariff provisions by the ICC would invite cases in more sensitive areas, Prairie brought a parade of witnesses to support the 100,000-barrel minimum. Despite Prairie's efforts, ICC ordered Prairie to reduce its minimum tender to 10,000 barrels. The order was not a blanket prohibition against unduly large batch sizes, however, and ultimately had little effect on oil pipelines in general. Prairie's actions demonstrated to independent shippers thereafter that any such formal complaint to the ICC by shippers would meet with an outsized and stubborn response from the pipelines. There would be no other material case brought by shippers against the integrated oil pipelines in the pre–consent decree era.

After the consent decree, the oil pipeline industry continued to develop with a high degree of vertical integration and joint ventures. At about the time when the ICC was dissolved and jurisdiction passed to the FERC for regulating oil pipelines, the DOJ lamented the interlocked shipper-ownership of the oil pipeline system and resulting lack of strong contending constituencies that could mount effective challenges to any aspect of pipeline rates, regulatory rules, or expansion plans—the DOJ's observation about the "absence of adverseness."

An unusual tariff case for oil pipelines was heard by the FERC in 2006 and 2007 involving the Trans Alaska Pipeline System (TAPS), which carries oil from Alaska's North Slope eight hundred miles south to the port of Valdez. The case represents an uncommon example of collective action against a vertically integrated, joint venture pipeline.[49] In that case, two independent Alaska North Slope producers, Anadarko Petroleum Corporation and Tesoro Corporation, joined the state of Alaska in opposing the rates proposed by the TAPS owners.[50] Those two producers and Alaska prevailed against the TAPS owners (representing a joint venture of the major integrated oil producers on the North Slope) in an extensive proceeding before the FERC, effectively halving the per-barrel TAPS tariffs. The effects of the ruling were to lower tariffs for Anadarko/Tesoro, which were not partners in the line, and increase the North Slope royalties for Alaska.[51] Had Anadarko/Tesoro been partners in the pipeline and thus been unwilling to challenge the rates in open administrative courtrooms, it is questionable whether Alaska alone could have prevailed.

CHAPTER SEVEN

Transacting with Private Carriage: The Gas Pipeline Regulations of 1938

In the first decade of the twenty-first century, whenever a question about competitive gas prices arises anywhere in the world, it is never long before the discussion turns to the current price at the Henry Hub in Erath, Louisiana. The hub facilities themselves are owned by Sabine Pipe Line LLC, and they connect to nine interstate and four intrastate gas pipelines. It is the physical pricing point for natural gas futures traded on the New York Mercantile Exchange (NYMEX). LNG tanker prices on the high seas, complex price-setting arbitrations in Europe and Australia, and Russian studies of the value of east Siberian gas—still in the ground and someday destined for Beijing and Shanghai—all eventually reference the Henry Hub, even if it is half a world away and hundreds of miles from open water. Why? The reason is that the Henry Hub is the center of the vigorous gas market in North America—the only competitive gas commodity market in the world to exhibit competitive spot and futures trading.[1]

The Henry Hub rose to prominence only as the US gas pipeline system evolved to support a market in legal entitlements to transport gas—an evolution that essentially culminated in 2000. Vigorous web-based trading occurs in point-to-point transport capacity rights (which are the well-defined entitlements that owners can freely buy and sell). Vigorous competition occurs for the building of new capacity (which essentially includes the making of new entitlements in point-to-point capacity). Competition for pipeline use and expansion exists on a pipeline system with utter transparency on cost and capacities. And yet, the pipeline owners themselves charge regulated prices. Captive pipeline users pay cost-based tariffs, new capacity must be federally licensed based on a traditional "economic need" regulatory test, and the regulators take pains to assure that buyers and sellers of gas transport remain fully informed.

Sixty-five years before, in 1935, the already extensive US gas pipeline system was highly vertically integrated into multistate utility holding companies. Those holding companies were closed books to outsiders, unregulated by the federal government, and widely considered a public scandal due to their anticompetitive and acquisitive practices combined with highly risky financial engineering. Somehow, contrary to the persistent vertical integration in the oil pipeline industry, gas pipeline transport transformed over the course of sixty-five years into an industry that exhibits true Coasian bargaining in transport entitlements and supports the world's only vigorously competitive and openly transparent gas market with an equally vigorous futures market.

Unregulated Transacting: Vertical Integration of Gas Pipelines Leading to the Holding Company Act of 1935

In 1936, Emery Troxel examined the costs and organization of long-distance gas pipelines. He noted that in addition to a lack of any federal price or service regulation, no federal or state agency gathered data on gas pipeline construction. Nor were there accepted points of separation between gas transmission, local distribution, or field gathering lines. Nevertheless, using an assortment of publicly available data from industry directories and investment advisory services, he found that more than 60 percent of "trunk" gas pipelines were controlled by the nation's five largest utility holding companies.[2] Other industry analysts found that by 1935, almost 80 percent of the gas pipeline mileage in the United States was part of nine major holding companies' systems, with extensive vertically integrated holdings in both gas production and distribution.[3]

Investment analysts at the time viewed vertical integration as a strength of the business, almost to the point of precluding a discussion of any other form of organization.[4] The holding company structure was driven in part by the mutual dependence between gas pipelines and producers and local distributors and reinforced by the need to form complex industrial relationships in an era when, as Troxel found, it was hard to obtain industrial information for such infrastructure businesses. Such was particularly true for interstate gas pipelines, which existed in a regulatory vacuum, with no agency in charge of collecting or publishing data.

The holding company structure adopted by electric and gas utilities in the United States during the 1920s and 1930s enabled a number of abuses, including writing up subsidiary property values and charging excessive service fees through affiliates.[5] The abuses were a highly public and political affair, particularly after Samuel Insull's utility holding company empire famously collapsed in late 1931 and early 1932—not unlike the controversy surrounding the unexpected collapse of Enron seventy years later.[6] The holding companies' primary abuse of power involved pyramiding control over regulated franchises: they allowed excessive returns at the top, in conjunction with extraordinary risk of financial collapse at the bottom with even the slightest nonperformance by the regulated franchises. No federal regulation of electric or gas companies existed prior to the 1930s, and state commissions could not effectively regulate the organization of holding companies. Many state commissions did not have statutory control over property and security acquisitions, mergers, and consolidations. Some commissions had indirect regulatory authority to oversee holding company activities in these areas, but the commissions did not exercise these powers. Troxel stated, "[State commissions] did not lack regulatory powers so much as they lacked perception, a strong feeling of public responsibility, and vigor."[7]

In February 1928, the Senate asked the Federal Trade Commission (FTC) to conduct an investigation of the public utility holding companies. The FTC produced a comprehensive and massive report in 1934 and 1935, ultimately comprising ninety-six volumes. The report showed that over half the gas produced and more than three-fourths of the interstate pipeline mileage in the United States were controlled by eleven holding companies. The four largest holding companies controlled 58 percent of the pipeline mileage. The holding companies had also branched out into manufactured gas, electricity, oil production, and coal. The FTC report highlighted many gas market abuses perpetrated by the holding companies, including monopolistic control of gas-producing areas, unreasonable differences in city "gate" (i.e., wholesale) gas prices, pyramiding investment schemes in gas enterprises, excessive profits on transactions between affiliates, inflation of assets and stock watering, and misrepresentation of financial conditions.[8]

Congress dealt with the abusive market behavior of the holding companies by passing the Public Utility Act in 1935.[9] Title I of that act (known as the Public Utility Holding Company Act, or PUHCA) gave the Securities and Exchange Commission (SEC) jurisdiction over public utility securities. As part of its new jurisdiction, the SEC was given the power to simplify the holding company structures. The SEC began by inviting the holding companies to submit voluntary proposals for their own reorganization; if each company could not justify its continued existence by 1940, then the SEC would initiate formal dissolution procedures.[10] The SEC's goal was to establish integrated distribution systems that were confined to a single regional area and to ensure that no holding company was so large as to impair local management, effective operation, or effective regulation.[11] In the end, the corporate distinction between gas pipelines and local distributors, which had blurred within a number of holding companies, became sharply defined.

Troxel minced no words in describing the unusually severe nature of the Holding Company Act ("the most stringent, corrective legislation that ever was enacted against an American industry . . . [a] remedy well suited to the patient").[12] He saw the ensuing strong enforcement role of the SEC as a just remedy for the abuses of the vertically integrated holding companies.[13] In passing the Holding Company Act, Congress generally ended the vertical integration of gas pipelines and gas distributors. As a result, the amount of interstate gas pipeline mileage controlled by holding companies declined from 80 percent in 1935 to 18 percent in 1952.[14] To use the term currently popular in Europe, it was the most comprehensive and forceful *corporate unbundling* of pipelines in history. The subsequent relationship among companies holding extensive relationship-specific investments became almost purely contractual. There was no standard form of gas pipeline/distributor contract in existence in 1935. But by the time the SEC began enforcing the act in the 1940s, Congress provided a standard contract (in the form of terms and conditions of regulated pipeline tariffs) as part of the Natural Gas Act (NGA) of 1938, the eventual outgrowth of Title III of the original Public Utility Act of 1935.

Like the Hepburn Amendment before it, the Holding Company Act was a stringent piece of legislation, through the passage of which Congress overcame its normal aversion to dealing with the complex internal structure of US corporations. The Hepburn Amendment in its time swept through Congress, borne along by the public's reaction to the Standard Oil Company's exploitative practices, a response that the Theodore Roosevelt administration carefully orchestrated in order to ensure wide-ranging ICC regulation of the oil industry. The Public Utility Holding Company Act, however, was a much more invasive piece of legislation, as it prescribed an unprecedented structural reorganization of the US utilities. Informed by the massive FTC investigation, Congress was willing to allow the SEC to restructure the internal workings of a major infrastructure industry. It was the last time that Congress was willing to bypass widespread industry opposition to take drastic action regarding the corporate structure of interstate pipelines.

The Natural Gas Act of 1938

The Natural Gas Act of 1938 was an unusual piece of legislation. Distinctly unlike the Hepburn Amendment of 1906, it specifically avoided the common carriage, railroad-inspired model of regulation. Instead, Congress gave the Federal Power Commission (the FPC—precursor of the FERC) the authority to regulate interstate gas pipelines as public utilities: specifying accounting practices, licensing new lines, and facilitating a priority of service to those local distributors that constituted the dominant pipeline customers. The federal government's switch in regulatory methods was influenced by the experience of the states in regulating their own utilities.[15] Equally important to the development of the new generation of pipelines built in the late 1940s, however, were various accounting, administrative, and constitutional institutions that formed the bedrock of future American utility regulation, all of which evolved during the moratorium on gas pipeline construction during the Great Depression and World War II.

Legislative Development of the Act

The first draft of the Holding Company Act was cosponsored by Representative Sam Rayburn (D-Texas), whose committee had asked for the investigation into the holding companies.[16] Introduced early in 1935, the draft contained three parts: Title I concerned holding companies (which went on to become the Holding Company Act later that year); Title II concerned federal regulation over interstate transmission of electricity;[17] and Title III proposed common carriage for the interstate movement of natural gas. Title III was not included in the final form of the act. There is no specific reason recorded in the Senate floor debate for why Title III was omitted, but it is clear that it lacked a strong constituent backing.[18]

Four points in Title III raised major objections from gas pipeline companies: (1) common carriage status and the associated obligations, (2) a blanket provision requiring new pipeline licensing, (3) the cost basis for determining regulated pipeline tariffs, and (4) the regulation of gas sales by gas pipeline companies to industrial customers. With common carriage, the pipeline companies did not see how they could reliably serve local distributors and other gas users if they were obligated to serve all comers. Indeed, the FTC, in its report to Congress, recognized that the gas industry was structurally different from other common carriers. The gas pipeline industry, aided by the FTC report, objected strongly to common carrier status, which it saw as inconsistent with its operations and commitments to gas customers—particularly gas distributors with their masses of residential and commercial customers.

Regarding Title III's blanket licensing requirement, the gas pipeline companies protested that the historical purpose of licensing was to limit competition in order to support the provision of efficient transportation services. Under this rationale, they argued, licenses should only be required for pipelines proposing to serve a market already served by an incumbent pipeline supplier. They also objected to the historical cost-based proposal of the FTC. Oil pipelines were then regulated by the ICC according to a "valuation" standard that placed less restrictive constraints on pipeline prices than those that might be imposed under a strict cost standard.[19] Finally, the gas pipeline companies saw the industrial gas sales market as highly competitive. They saw no need for any price regulation on gas sales to that market. Title III of the 1935 bill obviously involved controversial subjects and sparked legitimate industry concern. It is unsurprising, then, that the Senate removed Title III, whose controversial nature could have inhibited the urgent legislative action against the holding companies that was contained in Titles I and II.

In May 1936, Representative Clarence F. Lea, a Republican from California and the Commerce Committee chairman, introduced a bill similar to Title III of the original holding company bill. Lea's bill outlined the regulation of high-pressure interstate gas pipelines by the FTC. The Lea bill proposed again that the FTC be granted the power to fix just and reasonable rates or to eliminate any unreasonable rate differences for all high-pressure gas pipelines in interstate commerce. Under Lea's bill, the FTC would determine the cost of gas pipeline service based on an investigation of the "actual legitimate cost" of any high-pressure gas pipeline, including the original cost of all property used in gas pipeline service.[20] According to the bill, a gas pipeline company could be required to extend service to communities "immediately" adjacent to the pipeline but could not enlarge facilities in a way that would impair service to existing customers.[21]

Bowing to the concerns of state regulators, the Lea bill included a specific provision that denied the FTC the power to regulate sales of natural gas from low-pressure mains. As a result, the bill garnered the support of state regulators who had previously opposed the old Title III in the 1935 bill. The Lea bill, however, did not address the licensing issue that was of major concern to pipeline interests, the right to construct gas pipelines, or any specific financial controls of major concern to shippers and state regulators. Congress adjourned in 1936 before any of these issues were resolved.

When Congress reconvened in 1937, two similar bills came to the floor of the respective houses of Congress, both of which revolved around regulating gas pipelines as common carriers. They assigned common carrier status to gas pipelines if the lines carried gas for third parties and prohibited any undue price discrimination in the gas fields if the pipelines purchased gas to resell to distributors and others. The bills also obligated pipelines to serve gas to any community willing to extend a service line to the main pipeline, a requirement that reinforced common carrier status. They were strongly supported by the Cities Alliance, a group of one hundred midwestern city and town governments, which had organized in the mid-1930s to lobby for gas pipeline regulation, and which was vigorously opposed by pipeline companies. In addition to the common carriage obligation to connect new users, which could impair commitments to existing pipeline customers, the bills did not contain a provision that would limit rivalry among competing pipelines.[22]

In response to the dispute over the common carrier and licensing provisions of these bills, Representative Lea reintroduced a significantly altered version of his earlier bill in January 1937.[23] The new Lea bill placed regulatory responsibility with the FPC rather than the Federal Trade Commission (FTC), perhaps acknowledging that agencies charged with competitive market monitoring and antitrust enforcement are inherently poorly suited to the task of regulating franchised monopolies. Second, it exempted end-use industrial sales (as opposed to sales-for-resale gas distributor sales) from the jurisdiction of federal regulators, thereby removing one of the pipeline companies' key complaints. Third, it included a modified section on accounting regulation and cost determination, which the pipeline companies did not find particularly objectionable. Fourth, and perhaps most significant, the bill included a new component, Section 7(c), which limited the need for an FPC license to those cases in which a pipeline would serve a market already served by an existing pipeline, a provision ostensibly protecting interstate pipelines from pipeline-to-pipeline competition.

Despite the effort to bridge the gap between the needs of the pipeline companies and the demands of the Cities Alliance, the new Lea bill caused problems between the two groups. The pipeline industry liked the new provisions, particularly the new Section 7(c) relating to licensing of pipelines entering markets served by an incumbent, and gave its support to the bill. The industry's spokesman, W. A. Dougherty, an attorney from New York, said, "We think that generally it is sound regulation."[24] The Cities Alliance, however, wanted both to cap the price of gas delivered to cities and to foster competition among pipelines in order to lower prices and provide better service. Lea disagreed, arguing that gas producers and consumers could not have it both ways. He held that regulation by the FPC was better than an unregulated marketplace for gas distribution, saying, "That is what regulation is, monopoly controlled in the public interest."[25] Lea's point of view prevailed, and the Natural Gas Act passed the House and Senate in June 1938 without other major controversies or amendments.

Key Distinguishing Provisions of the Act

Most of the Natural Gas Act's twenty-three sections dealt with matters of procedure, law, penalties, and other issues that were unrelated to how the pipeline industry would transact with producers and shippers. But there are a handful of sections in the act that distinguish it from any other federal regulation of inland transportation. Some of the sections satisfied major constituencies, like the states or the pipeline industry. Others reflected three decades of advances in utility regulation since Congress had last dealt with pipeline regulatory legislation with the Hepburn Amendment in 1906.

LIMITING FEDERAL JURISDICTION TO SATISFY THE STATES.   Section 1(a) identified the gas pipeline industry as a matter of public interest—that is, it asserted the need to regulate the industry in some fashion.[26] Congress made sure to define the limits of the act by saying, in Section 1(b), that it "shall not apply . . . to the local distribution of natural gas or to the facilities used for such distribution or to the production or gathering of natural gas."[27]

REJECTING COMMON CARRIAGE TO SATISFY EXISTING GAS PIPELINE USERS.   Common carriage continued to be a subject of intense interest. Congress had debated whether to regulate gas pipelines as common carriers in the same way as railroads and oil pipelines. This time, the act explicitly rejected common carriage obligations in favor of private carriage for gas pipelines.[28] Section 7(a) stated that the commission "shall have no authority to compel the enlargement of transportation facilities for such purposes, or to compel such natural-gas company to establish physical connection or sell natural gas when to do so would impair its ability to render adequate service to its customers."[29] In essence, the commission would not give existing and new pipeline customers equal priority. The gas pipeline companies' commitment to existing customers had to come first. The requirement was the antithesis of traditional notions of common carriage or nondiscriminatory third-party access (TPA), in which all shippers are served and charged equally.

The importance of this provision in the future development of competitive US inland gas transport has been monumental. This was just the sort of provision that Senator Joseph Foraker would have admired in an earlier Congress: he argued in 1906 that pipelines were not in fact common carriers. In the context of the Hepburn bill in 1906, however, it would have been seen as an unthinkable concession to Standard Oil given that company's prebreakup dominance of the US oil business. Coming on the heels of the Holding Company Act, however, this provision confirming private carriage for interstate pipelines was merely a practical recognition of the need of gas distributors and others to count on uninterrupted supply in the pipelines that were constructed to serve them—and whose millions of connected customers ultimately provided the creditworthiness necessary to motivate the provision of the capital involved.

LIMITING ENTRY TO SATISFY INCUMBENT PIPELINES.   The next critical economic feature of the bill concerned conditions of entry and possible pipeline rivalry. Consistent with Representative Lea's philosophy that regulation is "monopoly controlled in the public interest," Section 7(c) stated that the FPC would judge the economic need of any interstate gas pipeline proposing to enter a market already served by another.[30] This provision appears to favor incumbent pipelines in existing markets based on the fact that they can presumably serve—and continue to serve—at a lower cost than a new pipeline.

INVOKING THE JUST AND REASONABLE RATE STANDARD.   The final point of contention in the Natural Gas Act concerned the setting of prices. Congress applied the ancient "just and reasonable" standard from the common law.[31] While the just and reasonable standard may seem ambiguous to the layperson or to those in Roman/civil law countries, it has a long-accepted legal interpretation under the common law. The provision gave the FPC full power to investigate and adjudicate what it considered the fairness of the rates charged by interstate gas pipeline companies.

REGULATING ACCOUNTING.   With more than three decades of experience since its 1906 Hepburn Amendment, and with full knowledge of the ICC's failure to settle accounting matters on its own with respect to oil pipelines, Congress invoked accounting regulation at the federal regulatory level for the first time. Section 8(a) of the Natural Gas Act gave the FPC the ability to control accounts for rate-making purposes.[32] The entire history of oil pipeline regulation by the ICC was hindered by accounting ambiguity in common carrier charges.[33] Congress had also just dealt with the holding companies, whose abusive financial accounting practices, among other things, led Congress to give the SEC unprecedented powers to rectify the problems. Congress had learned that weak accounting regulations enabled abusive practices on the part of regulated companies.[34] It took the FPC only two years to create a regulatory accounting standard tailored to just and reasonable rate making, which became the general model for all subsequent state regulatory accounting statutes in the United States.[35]

OTHER PROVISIONS.   The Natural Gas Act also contained provisions concerning the abandonment of lines (Section 7[b]), regulation of depreciation practices (Section 9[a]), rules pertaining to administrative procedures (Section 15[a]), procedures for rehearing and appeal of commission orders (Section 19[a]), and issues pertaining to the FPC's enforcement powers (Section 20[a]). In all, the act provided an effective framework for regulating price and entry for interstate gas pipelines. It resolved the issues raised by the state commissions, the gas pipeline company interests, and pipeline customers, and, importantly, it relied on the quasi-judicial powers of the FPC to deal with issues arising from the collision of interests among pipelines, their customers, and the public interest.

US Legislative Politics and the Avoidance of Common Carriage

The Natural Gas Act came about in response to pressure from the states and gas-consuming city coalitions, fueled by the FTC's ninety-six-volume investigation. Their main concern was the state regulatory commissions' inability to deal effectively with the possible exercise of market power over the price of gas delivered to distribution companies. Critically, however, the act had also been shaped by Congress to deal with pipeline company concerns that common carriage was an unworkable transactional structure for gas pipelines and their feeling that incumbent pipelines serving established markets should be protected from rivalry. The easiest way for Congress to deal with these concerns was to craft a regulatory formula that acknowledged the status quo industrial structure. Essentially, Congress left the basic structural organization of the industry in place (post–Holding Company Act of 1935) while giving the FPC the authority to investigate, impose standard accounting practices, judge just and reasonable rates, and license new entrants. This was standard utility-style regulation at the time. As the act's sponsor, Representative Lea stated that this was "regulation along recognized and more or less standardized lines. There is nothing novel in its provisions."[36] Representative Lea knew what he was talking about—he had been a member of the Public Utility Commission of California before his tenure in Congress.[37]

Some modern writers have treated the imposition of gas pipeline regulation on the utility model, rather than the common carrier model, as a missed opportunity.[38] But it is hard to agree, given the industry that Congress had to work with and the methods Congress used for such legislation. Except in truly extraordinary situations, like the 1935 Holding Company Act, Congress generally avoids proposing major structural reforms of existing industries. In the 1930s, Congress was unwilling to pursue new pipeline industry structures based on a concept that had only been applied to railroads, particularly since the pipeline industry during the Great Depression was performing poorly. Further, there was never any prospect that Congress or the Franklin Roosevelt administration (which included Thurman Arnold) would have let gas pipelines follow in the vertically integrated footsteps of oil pipelines.

Validation by the Courts of the Natural Gas Act: The *Hope Natural Gas* Case and the Valuation of Property under Regulation

Unlike regulations in most other places in the world, particularly outside of common-law countries, major regulatory statutes in the United States do not become settled methods of government control over private businesses until they are tested in the courts. In this respect, the US Supreme Court is the ultimate regulator. The major test for the court revolves around *property*—that is, whether the new regulations somehow deprive investors of the value of their property without due process of law (and due compensation) under the US Constitution. The test case for the Natural Gas Act in the courts settled the question of the value of regulated utility property with such finality that it has long since ceased to be a matter of research for US economists.

The last chapter showed how it took eight years for the Supreme Court to overcome the objections of the oil pipeline industry (largely funded by Standard Oil) regarding the legality of the ICC's tariff-setting authority. In a remarkably parallel story, again funded by Standard Oil (through a gas pipeline subsidiary in Ohio) but taking only six years this time, the Supreme Court again confirmed the FPC's jurisdiction over pipeline rates consistent with the US Constitution. This second time around, the court dealt with the question in such a general and well-founded way, particularly with respect to how to value regulated property, that it established precedent for all regulated enterprises in the United States into the twenty-first century and beyond.

The Supreme Court itself had flip-flopped, over the course of half a century, on the subject of whence the value of that property comes for rate making. In an 1898 decision (*Smyth v. Ames*), the court ruled that the valuation of utility property used in rate making should be based on "fair value."[39] This decision, which stood for almost fifty years, created difficulty for regulators because the term *fair value* had no objective meaning. In practice, the *Smyth* decision entailed a compromise between original cost and "reproduction cost" foundation for tariffs. Troxel criticized the ruling, saying that fair value was "only a 'soothing phrase,' a way to cover up the difficulties of reasonable property valuation and to avoid future commitments."[40]

In 1940, Ben W. Lewis, of Oberlin College, colorfully summed up the era of "fair value" in rate making, calling it little short of a public scandal.[41] The scandal ended when, in 1944, the Supreme Court ruled in *Federal Power Commission et al. v. Hope Natural Gas Co.* The Hope Natural Gas Company was a Standard Oil Company gas pipeline subsidiary that filed suit against the FPC over its first ruling under the Natural Gas Act of 1938. With the *Hope* ruling, the Supreme Court set a new standard for determining "just and reasonable" returns for investor-owned utilities: "The return to the equity owner should be commensurate with returns on investments in other enterprises having corresponding risks. That return, moreover, should be sufficient to assure confidence in the financial integrity of the enterprise, so as to maintain its credit and attract capital."[42]

In setting permissible revenues, a utility's profit (resting on invested capital as reflected in accurate bookkeeping) would be measured by potential earnings for investors based on other enterprises of similar risk. The *Hope* decision secured utility companies' investments from seizure (a "taking" of private property without due process) if regulators set charges to award returns consistent with investors' opportunity cost of the equity capital invested in regulated enterprises—as recorded in those enterprises' books. The decision cut through the uncertainty that had dogged the value of capital in regulated business for decades and set the stage for a predictable cost basis for shippers' transport entitlements that would later form the Coasian gas transport market.

The *Hope* decision produced what is known as the "end result" doctrine, which flipped the "just and reasonable" requirement on its head: the ultimate effect of price control on investors' property had to be determined as just and reasonable, not the method for achieving the result. Supreme Court Justice William O. Douglas, who wrote the majority opinion, said, "It is the result reached, not the method employed, which is controlling. It is not theory but the impact of the rate order which counts. If the total effect of the rate order cannot be said to be unjust and unreasonable, judicial inquiry under the Act is at an end."[43] The *Hope* decision was otherwise unremarkable in the way it progressed through the legal system. It was a typical event in the US common-law system, in which the Supreme Court took up a case brought by an individual or corporation against the state or federal authorities on the grounds that it conflicted with the Constitution's protections of property.

The *Hope* decision was a landmark event in the history of the economics of utility regulation. In the context of a well-structured Natural Gas Act resting on a legislatively mandated Uniform System of Accounts, it meant that federal tariff regulation for pipelines was to become a highly reliable and predictable affair. Those economists who lived through the issues leading to that judgment vied with one another to memorialize its importance. When the ink was barely dry on the *Hope* case, various economists waxed poetic on that decision's profound effect on rate regulation. Troxel's contribution emphasized the aimlessness of investigations of *regulated value* rather than *compensatory earnings*.[44] Eli Clemens, of the University of Maryland, went further in his praise for the *Hope* decision, describing it as a fundamental advance in the theory of administered value in law.[45] James Bonbright, of Columbia University and already the recognized expert among scholars of the theory and practice of property valuation, called it "one of the most important economic pronouncements in the history of American law."[46]

Bonbright's statement was not hyperbole. Throughout the 1920s and 1930s, economists and legal scholars had struggled with the question of how to value regulated property—even to the point of speculating that public ownership might be the only way out of the difficulties arising from the lack of any meaningful economic definition of "fair value" for the property of regulated public utilities.[47] Bonbright led the research on that question, with many named collaborators in economics, law, accounting, and finance, under the auspices of the Columbia University Council for Research in the Social Sciences. The culmination of those efforts, his massive 1937 two-volume treatise *The Valuation of Property*, made Bonbright the obvious valuation witness for the FPC in the regulatory litigation leading to the *Hope* decision. With that decision, the basic theory of value for regulated US utility property became a settled issue. By 1960, Bonbright would write that "all experts" generally agreed that the valuation of property as part of the formula for permissible regulated revenues was distinct from valuations in nonregulated spheres, and that the reasons why it took the Supreme Court half a century to come to that conclusion were of "historical interest only."[48]

On this last point, Bonbright would be proved wrong. Only econo-

mists in North America were convinced—and then only those who had
lived through the shift from reasonable valuations to reasonable earnings on invested capital as a component of regulated tariffs.[49] When the
pipeline privatizations arrived in the late twentieth century, the research
leading to the *Hope* decision was largely forgotten. In both the United
Kingdom and Australia, basic utility regulation looked once again to
property valuations rather than reasonable earnings on invested capital
as the source of tariffs. Argentina made a halfhearted effort to use book
capital as the basis for reasonable earnings, but that effort waned and
ultimately was mooted by that country's noteworthy expropriations in
early 2002.

The Problems with the Public Utility Model for Pipeline Transacting

The Natural Gas Act had rejected common carriage and had called
for pipelines to transact as utilities—as full-service providers for local-
distribution gas companies at their "city gates" (where they connected to
the interstate pipelines). Interstate gas pipeline companies would build
and operate pipelines, operate storage facilities, and buy gas supplies in
the field to be resold to gas distributors and others at cost. Such utility
regulation had worked well enough for energy distribution monopolies.
But interstate pipelines were competitive rivals in seeking FPC licenses
for new lines to satisfy the rapid postwar demand for gas. It was the race
by pipeline companies to secure blocks of gas to fill those prospective
lines, with an eye toward winning licenses, that caused gas distributors
and the gas-consuming states to complain that pipelines were not sensitive to the quickly rising price of gas in the field. The representatives
of gas consumers felt that pipeline companies, acting as gas buyers but
limited to earning investment returns only on transport, were heavily
skewed away from the kind of price sensitivity that would be expected in
unregulated markets.

The consuming states (and their gas distributors) convinced both the
Supreme Court and the FPC, over the objections of gas producers, of the
unacceptable consequences of allowing pipeline companies to bargain
directly with gas producers in the field for gas that they would resell to
distant and largely captive customers. The Supreme Court in 1954 thus
directed the FPC to regulate wellhead gas prices, and the FPC in 1965

further chose a complicated two-tier method proposed by those distributors. The period saw almost nonstop litigation over the issue and strong calls from a large portion of the community of economists for Congress to pass new legislation deregulating gas prices or for the FPC to set higher prices in the field.

Having pipelines buy the gas in an unregulated market, in a rush to obtain new pipeline licenses, was a genuine problem that the courts and the FPC could not ignore. But simply regulating prices was no long-term remedy either, as that would merely pit gas producers and distributors (and their respective economists) against each other in a decades-long administrative battle that had no real prospect of keeping up with a complex and inherently volatile market for such a fuel. Ultimately—perhaps inevitably—the pipeline companies as a group got into financial trouble, allowing the FERC to extract "voluntary" open access from pipelines by the 1990s via a partial industry bailout.

Those who spent years litigating these issues before the FPC, the courts, and ultimately the FERC tend to look upon these decades as packed with important litigation surrounding a very complicated set of developing issues. That complexity is a mystery to most outsiders and is tedious to explain in detail. But the basic elements of the story are simple enough: semirival regulated pipeline companies, which profit only through a return on their regulated transport investments, simply do not make responsible agents for the purchase of gas on behalf of captive distribution utilities, in illiquid markets dominated by long-term contracts. Trouble was inevitable (shortages, surpluses, heavy litigation, financial failure, etc.) until those distributors could buy gas in the field for themselves and simply contract with pipelines for the needed inland transport service.

This is not to fault Representative Lea for imposing a defective regulatory transacting scheme on the industry in the Natural Gas Act. By the 1930s, it was evident that common carriage was ill suited to an independent pipeline industry, and the utility model was the only known alternative. There was no experience with the kind of transacting needed for a transparent and competitive pipeline transport business—such a market had not even yet been described by Ronald Coase to a highly skeptical economics profession. That experience would come only after the hard-fought battles among producers, pipelines, and distributors from the mid-1950s right up until the end of the century.

Stripped of the play-by-play drama of the events as they unfolded during those years, four milestones signify the transition from regulated private carriage in 1938 to contract-based competitive pipeline transport in 2000. First was the Supreme Court's refusal to interpret the Natural Gas Act as allowing pipelines the freedom to contract for gas supplies for captive customers without regulatory oversight. Second was the long battle between the producers and distributors on whether to deregulate those pipeline gas purchases—a battle that the distributors won. Third was the trouble that pipeline companies, as gas buyers, faced during the turbulent energy markets of the 1970s and 1980s as Congress tried haltingly to relax price controls. Fourth was the partial bailout of the pipelines that led them "voluntarily" to transform their gas sales obligations to gas distributors into open-access transport contracts.

The Supreme Court Orders the FPC to Regulate the Wellhead Price of Gas

From the perspective of the economics of public choice, the Natural Gas Act reflected a pluralist solution to a set of political exigencies that had already largely passed into history by a decade after its passage. The act indeed set the stage for highly effective cost-based regulation of interstate gas pipelines that works well today. But the act did nothing to resolve the problem of leaving a limited number of pipeline companies doing all the gas buying for captive gas distribution utilities in the race for additional pipeline licenses. Such fundamental industry regulatory legislation is durable in the United States—and indeed, the Natural Gas Act survives largely intact in the twenty-first century. It took the courts to isolate the act's deficiencies in the market and set the stage for the battles among adverse interest groups that led to the development of competitive Coasian gas transport decades later.

Given the action Congress took in the 1930s to stem pricing exploitation by the utility holding companies, it should not have been surprising that the Supreme Court would interpret the Natural Gas Act as extending FPC jurisdiction over gas sales to companies affiliated with regulated interstate pipelines. What the Supreme Court did in its next major gas case, however, was more radical. In the landmark *Phillips Petroleum Co. v. Wisconsin*, the court directed the FPC to regulate all gas prices, even those at arm's length between pipelines and producers. Arguing before the Supreme Court, Phillips Petroleum tried to contend that only pipeline regulation was the subject of the Natural Gas Act. The court disagreed, noting that if this had been Congress's sole intent, it would not have made mention in Section 1(b) of "the transportation in interstate commerce" and separately of "the sale in interstate commerce of natural gas for resale."[50] The case was remanded to the FPC for the regulation of the prices of all gas sold to interstate pipelines, thereby sparking forty years of controversy.

The *Phillips* decision came under withering criticism later from most economists and legal scholars when it became clear that the administration of federal price regulation was incapable of keeping up with supply and demand in the relatively volatile and uncertain market for gas. Richard Pierce called the opinion "infamous."[51] Others took a less disparaging view. Stephen Breyer and Paul MacAvoy called the Supreme Court's logic "not wholly unreasonable."[52] Much of the criticism of the *Phillips* decision in any event was myopic, as it failed to consider what problems would have attended the unregulated sale of gas under long-term contracts to pipelines acting as agents for captive gas distributors. When Representative Lea, the former California regulator, sponsored the Natural Gas Act based on utility regulatory principles in existence at the time, it would have been unreasonable not to give Congress the power to look at both transport and gas costs, since those gas pipelines sold *delivered gas*. Unless Congress was willing to split the gas market cleanly from gas transport, a move that was more than fifty years away, it could not have written regulations denying the FPC the ability to look into all pipeline purchases—whether for steel, labor, or gas supplies. For the steel or labor, well-functioning markets existed upon which the FPC and others could judge the reasonableness of prices paid by the pipeline companies. Such was not so for gas in the field bought under long-term contracts. In this light, the Supreme Court's decision in the *Phillips* case was inherently reasonable and probably inevitable.

With a regulatory system that obliged gas distributors to buy from interstate pipelines—to which they would be captive once connected by the very nature of long-distance transport pipelines—*someone* had to judge the reasonableness of the prices those pipelines would pay. Indeed, the subsequent fights between gas distributors and producers over the issue showed just how suspect those transactions could be, especially as pipeline companies raced to acquire gas supplies to gain an edge over rivals in the veritable licensing sweepstakes before the FPC.

Licensing Rivalry, Distorted Gas Markets, and Pressure Groups

For local gas distributors, the licensing of new facilities is rarely contro-versial—as those distributors are the only gas suppliers in their respec-tive towns. Such is not true for long-distance pipelines, as the growing market for gas, and the depletion of gas fields, makes for constant shifts in the way gas ships from producing areas to markets. A number of ex-isting major pipeline companies engaged in long-distance gas transport when pipeline building began again in earnest in the late 1940s. It was up to the FPC to provide licenses for those companies to build new lines. Pipeline companies were originally required to obtain licenses only when they entered markets already served by another pipeline. But in 1942, Congress amended the Natural Gas Act to require licenses for all new construction, extensions, or acquisitions of gas pipelines.[53] The ri-valry to obtain those licenses thus became intense.

By the late 1950s, it was apparent to some economists that the pres-sure of licensing was affecting incentives in the bargaining for wholesale gas supplies. Joel Dirlam, of the University of Rhode Island, said in 1958 that the demand in the desirable gas fields must be quite inelastic be-cause of time pressure on the pipeline to accumulate the reserves neces-sary for licensing, the commitments of the distributors, and the unique-ness of natural gas as a cooking and heating fuel.[54] While testifying in one of the major cases stemming from the FPC's attempt to regulate gas prices in 1959, Alfred Kahn also identified the effect of licensing on gas pipelines attempting to enter new markets.[55] Gas pipelines requiring li-censes for new lines needed to demonstrate that they had secured ade-quate gas supplies. The problem, according to Kahn, was that the "up-ward thrust" of prices in the fields was driven by the necessity of finding uncommitted fields under the pressure of the race to obtain certification (i.e., a license).[56] To Kahn, the pressure to secure the new supplies neces-sary to license new lines gave producers a power in contract negotiations with pipeline companies that they otherwise would not have had. The FPC was convinced that federal pipeline licensing distorted the long-term contracting practices and prices in the gas fields.[57] The FPC imple-mented in 1965 the suggestion of the gas distributors, presented origi-nally in Kahn's 1959 testimony, to split "old" and "new" gas prices. The split was intended to give incentives to search for sufficient new gas sup-plies while limiting the prices (that is, the "economic rent" above cost) of gas already flowing.[58]

The gas producers, which had pushed very hard for effective well-head gas price decontrol, either in Congress or before the FPC, hated the FPC's solution. The producers' view of the market is stated by Paul MacAvoy, of Yale, a student of Morris Adelman at MIT, the witness for the producers and Kahn's major adversary in the gas rate proceedings at the FPC. MacAvoy modeled a loss of consumer surplus of twenty billion dollars over the ten-year period 1968–77, saying that the "straightforward rationale for regulation was that it generated benefits from lower prices."[59] The producers' view, however, ignored the root of the problem that led the Supreme Court and the FPC to regulate in the first place—that pipelines were simply agents for captive gas distributors with manifestly skewed incentives when it came to buying gas for resale at cost.

Such a market for gas was bound to be distorted when producers pocketed the gains from higher prices (and more advantageous quantity price adjustment and quantity clauses like most-favored-nation and take-or-pay clauses, which kept both prices and billed volumes up, even if actual "takes" declined) while pipelines simply passed the cost and take obligations to distributors. Those distributors, representing millions of consumers and backed in their efforts by their state regulators, would never have agreed to let pipelines bargain on their behalf without close limits on their behavior and regulatory scrutiny of the results. There was never any chance that those representing the gas-consuming public would willingly permit pipelines to transact in the fields with gas producers as if the distortions did not exist. Neoclassical economic models that purport to show what would have happened in an equilibrium gas market during that period without price controls are simply not realistic.

To be fair, however, it was also unrealistic to think that the kind of price regulation the FPC would impose in the gas fields, through its lengthy administrative procedures that only applied to *interstate* gas sales, would be sensitive to the needs of producers in a complex, capital-intensive, and volatile energy market. In addition, the two-tier price regulations presumed that the FPC had the ability to regulate economic rent associated with flowing gas supplies. Dirlam correctly identified the action as "rent regulation" in the case of old flowing gas supplies—"whether Mr. Getty shall buy a yacht (or a Jackson Pollock) or whether thousands of New Jersey commuters shall enjoy an extra evening 'on the town' in Manhattan once a year."[60] He and others knew that the attraction of regulating gas rents as a policy depended completely on whether the regulation would avoid harming production incentives. He was convinced

that the supply elasticity was steep, asserting that "it is extremely doubt-
ful whether any practically conceivable change in the field price of gas
could result in any faintly predictable corresponding change in its long
run supply."[61]

Any attempt to regulate the economic rent associated with flowing
gas had to confront the inelasticity of supply and the opportunity cost
of the producers themselves, many of whom produced gas jointly with
oil. The FPC never had the ability to discern the true cost of production, which was a subject of dispute that its adversarial evidentiary procedures could not practically address. In addition, the FPC never had
the ability to overcome producers' ability to hold supplies back in order to wait for the market, the FPC, or Congress to allow higher prices
for interstate gas shipments. In the end, the FPC lacked the tools to determine an unambiguous cost of gas or to effectively regulate gas quantities. That is to say, it did not have the practical ability after all to appropriate Mr. Getty's Jackson Pollock for the benefit of New Jersey gas
consumers.[62]

The Pipeline Companies Overextend

Regarding the dispute over regulating the wellhead price of gas, Kahn
said in 1960 that "after several years of swimming in this sea of debate,
I feel confident in no other conclusions than that the opposing forces
are stalemated, so far as economic arguments alone are concerned."[63]
From the perspective of the twenty-first century, the 1960s debate was
between two partially correct but inherently impractical neoclassical positions. Producers wanted deregulated gas prices, which public opinion
would not (and ultimately did not) permit given the evident distortion
in the gas market caused by the institutional constraint of pipeline regulations. Distributors wanted to cap gas prices in order to regulate economic rent in the gas field, which the FPC did not have the practical ability to do. Something bad was bound to happen in the gas market either
way. Either price controls would lead to a shortage, or price deregulation
would lead to excessive buying and a costly surplus. The shortage came
first, the surplus came in reaction.

By the early 1970s, the wholesale gas price regulation had reduced *in-
ter*state shipments (but not *intra*state shipments) and contributed to the
perception of a gas shortage in the gas-consuming states. As a result,
many large gas users and industrial customers in those states were un-

able to secure reliable gas supplies, which various writers have estimated cost consumers many billions of dollars in lost consumer welfare.[64] Some reacted to the problem by calling for quick deregulation, including former US treasury secretary William Simon.[65] On the other side of the debate, Kahn, then chairman of the New York Public Service Commission, testified before Congress, saying, "Deregulation of gas will surely raise price gouging opportunities; the probable windfall to the industry is a price monstrously out of proportion to the benefits that deregulation might be expected to bring. I regard simple deregulation in these circumstances as totally unthinkable, and I cannot bring myself to believe that Congress will be willing to enact it."[66]

Congress responded with legislation in 1978.[67] It perceived that the shortage had developed in response to the rigidly controlled wellhead gas prices and declining reserves of the early 1970s combined with increased oil prices following the 1973 OPEC oil embargo. The process Congress used for deregulation was both complicated and gradual: one-half of the gas supply was scheduled for deregulation in three phases, while the other half was never scheduled to be deregulated.[68]

From the perspective of the volatile energy markets of the twenty-first century, it is easy to see that such complicated congressional legislation was no lasting solution to the problem. The underlying institutional difficulty—gas pipelines buying for their captive distributors—was still there. So when the effect of the OPEC oil embargo lessened in the late 1970s and the prices rose in the fields according to the new legislation, the pipelines did what might have been expected for a business that profited only through transport: they bought gas in overabundance, fully expecting to resell it to distributors at cost. Pipeline companies bought not only at comparatively high prices but also with stringent obligations to take gas or pay for it anyway (called "take-or-pay" contracts). In response to the rise in interstate pipelines' prices, many interstate pipeline customers (particularly industrial customers) tried to avoid buying the now-expensive pipeline-owned gas. Instead, these customers pursued individual certificates for "transportation" of cheaper gas through the pipelines than the pipelines themselves were able to offer.[69] This caused the pipelines to act less frequently as gas merchants and more frequently as transporters of third-party supplies, which amplified the pipelines' difficulties by shrinking their captive gas markets even further. By 1986, those unmarketable supplies totaled approximately $11.7 billion, threatening pipeline companies' financial integrity.[70]

Bailout in Exchange for Contract Access

In October 1985, the FERC responded to the pipeline companies' distress by establishing a voluntary contract access transportation program.[71] The FERC gave gas pipelines two options.[72] A pipeline company could accept "open-access" status and transport independently owned gas on a first-come, first-served, nondiscriminatory basis under a preapproved, standard license and existing transport contracts. Alternatively, it could decline open-access status and exclude all independently owned gas from its line (precluding it from offering selective transportation service and risking the loss of its customers to less expensive alternative fuels or less expensive sources of gas). To the extent that the acceptance of contract-based open access and the loss of their traditionally captive gas customers would confront pipeline companies with unmarketable gas costs, the FERC eventually offered to allow pipeline companies to pass along half of those costs as a transport surcharge for all shippers.[73] Given what had transpired in the turbulent energy markets of the previous decade, the second choice would have posed immense risks for pipelines as gas merchants. All major US interstate pipeline companies quickly volunteered to adopt contract-based open access.

Consolidation of Open Access into Competition in Transport

Up to this point in the story, things moved rather slowly, if not quietly. After 1986, however, events moved fast on a number of fronts. By 2000, the gas pipeline industry had been transformed into a competitive business for both the daily use and construction of transport capacity. In that time, gas pipelines went from being the dominant buyers and sellers of gas in the United States to owning none of the gas they shipped in interstate commerce. What happened?

What happened was that the market for gas transport transformed into a highly liquid market for essentially perpetual legal entitlements to transport gas from one point to another.[74] In that market, gas pipelines own and operate the price-regulated facilities that support those entitlements to transport gas, but they do not own or control the entitlements themselves, nor do they possess any operational or financial information that is not an open book to those who would buy or sell those entitlements. The entitlements themselves are explicit in terms of the physical transport they cover, have a highly predictable cost basis for those who buy and sell them, do not expire for practical purposes, and trade almost without friction in standardized web-based exchanges on a daily basis. That is, the legal entitlements to well-defined pipeline transport rights are a competitively created and traded commodity. Such is, of course, an example of the Coase theorem at work—perhaps the best example of all for the way in which an efficient market in well-defined legal entitlements so replaced an unworkable gas market that had depended on pipelines as intermediaries.

It is worth remembering that it was Coase who saw how the definition of property rights can endow a resource with institutional scarcity in order to form the basis for trade and the elimination of the need for the government to restrain those who would wish to use a limited resource. The creation of such institutional scarcity and a corresponding market came in three parts to the gas transport industry in the United States. The first was the transformation of a generalized notion of regulated contract-based open access into an exacting specification of physical transport rights that could be traded without the operational discretion of the pipeline company. The second was the creation of a predictable cost basis for those rights that buyers and sellers could rely upon for as far as the eye could see, hence facilitating both short- and long-term trades. The third was the invention of a modern electronic trading and information system where buyers and sellers could transact with full information and very little cost, delay, or uncertainty. With such a market, the price of existing rights readily compares to the cost of expanding the institutionally scarce resource, vastly simplifying the regulatory burden of approving new capacity—as "economic need" is evident in the willingness of shippers to commit to pay for new capacity.

Creating Highly Specific Physical Gas Transport Rights

When the interstate gas pipelines as a group converted from the business of providing delivered gas for their customers to open access after 1985, the transport entitlements were not well defined. It was as if the pipelines had accepted the *principle* without knowing how to put that principle into practice. The acceptance of open access was a theory that had yet to be put to the test.

Problems came up immediately in using pipelines to transport third-party supplies. Pipeline companies had for decades dealt with their major customers—the gas distributors—according to rules in their tariffs designed around being the full-service gas supplier at distributors' city-gate stations. The pipelines did not have open-access tariff rules, and the ones they initially created did not treat pipeline-owned gas and third-party gas equally. Two symptoms of this inequality showed up in the commodity market. First, the FERC found that while total pipeline gas sales constituted only 18.8 percent in 1990–91 on a yearly basis, the to-tal for the winter period (November through March) jumped to 65.8 per-cent.[75] Evidently, pipeline company customers that preferred to purchase third-party gas throughout the year rushed back to pipeline-owned gas during the potentially stressful winter peak. Second, the FERC saw a persistent premium in the price for pipeline-owned gas in comparison to third-party gas.[76] In a number of cases filed after the 1985 switch to open access, pipelines tried to implement various charges for holding their gas ready for sale—charges that pipeline customers considered a signal of pipeline companies' ability to reserve the most premium transport ser-vices for themselves.[77]

In 1992, the FERC dealt with this problem in a singularly effective way. It directed pipeline companies to charge separately for the vari-ous pipeline services provided to shippers. That is, the FERC required that "no-notice" service, through which pipeline companies must de-liver up to shippers' maximum contract quantities without requiring prior shipper notice, had to be priced separately.[78] Also, in a brilliant move, the FERC directed pipeline gas-marketing affiliates to transfer ti-tle to gas sales at "pooling points" far upstream. Downstream of these pooling points, all gas would be owned by shippers.[79] With the change of title to gas supplies at the pooling points, any subtle or perceived ad-vantage enjoyed by pipelines' gas-marketing arms vanished without the FERC's having to require the restructure of pipeline companies or the creation of some sort of enforceable information barrier between pipe-line companies and affiliated gas marketers. It was, for all intents and purposes, a highly inventive and effective version of the "commodities clause" that Congress had wrestled with in 1906 when debating the Hep-burn Amendment. With the prices attached to services that allowed for greater pipeline flexibility, and with the use of pooling points, any real or implied advantage for pipeline-affiliate gas was impossible to maintain, and the price disparity between pipeline-affiliated gas and third-party gas evaporated.[80]

Both the FERC's 1985 and 1992 actions foresaw—and moved a long way toward—a new market in which shippers could use their pipeline capacity rights to offer services in the market for point-to-point transport in competition with the pipelines' offerings of uncommitted firm or interruptible capacity. But barriers remained in the pipeline transport market in the precise definition of contract holders' physical rights to transport service, gas balancing, and flexibility. The FERC acted again, in 2000, after extensive evidentiary hearings, which dealt with the detailed operational work of implementing the main provisions of what it ordered in 1992.[81] That action required pipeline companies to modify their scheduling procedures to eliminate existing disadvantages for "released capacity" (i.e., the sale of transport entitlements to others) relative to pipeline-controlled capacity, thus allowing released capacity to compete on a comparable basis with pipeline-owned capacity. That action also required pipeline companies to permit shippers to "segment" capacity for their own use or release. Segmenting broke up capacity into separate segments in a complete chain, to facilitate using some segments and selling the entitlements to others. It revised imbalance management and penalty provisions (again, after extensive evidentiary hearings), limiting penalty assessment to only those cases in which reliable evidence demonstrated they were needed to protect system reliability. Finally, it required that any operational restrictions on firm transport customers' use of their contract capacity entitlements—for themselves or to sell to others—required evidentiary justification related to safe and reliable pipeline operation. What firm shippers ultimately got was a well-defined and reliable definition of the physical parameters of their transport rights.

Creating a Predictable Cost Basis for Transport Capacity Rights

As it worked toward resolving the rights to capacity under gas transport contracts, the FERC had to deal with a number of disputes that would determine whether those rights would have a predictable cost basis. Two items could unpredictably alter the cost basis for those who would buy or sell transport entitlements. The first item was tariff design: whether the fixed cost of transport service would be reflected in fixed regulated pipeline prices. The second item concerned the cost basis for those entitlements: whether the inherent market value attached to entitlements would remain with the pipelines' contract holders or would be drawn upon by pipeline companies to subsidize the building of new capacity.

The FERC dealt with the tariff design issue by directing pipeline

companies to charge "straight-fixed-variable" (SFV) rates. Such a tariff resembled contract rental payments for the transport entitlements, as
they would be largely invariant to how much gas actually passed through
the pipeline.[82] SFV tariffs, as opposed to more volumetric tariffs that the
FERC had employed all through the 1970s and 1980s, simply made the
cost basis for the purchase and sale of entitlements easier to predict—
and hence facilitated their trade.

The second item was potentially much more damaging to the market for those entitlements. It dealt with whether the pipeline companies
could draw upon the inherent value in existing entitlements to subsidize
("roll-in") the cost of new capacity projects. It was an issue that pitted
pipeline companies against their firm shippers over who would control
the value of transport entitlements when the value of transport capacity
between two points in the gas market exceeded its cost-based regulated
rate. In most cases, the regulated rates were lower than the current value
of the capacity in the market—often very substantially so.

The FERC had traditionally struggled with the issue in the pre-open-
access era, approving both rolled-in and incremental charges proposed
by pipeline companies. In addition, before the era of open access, some
pipeline companies had systematically rolled in all new pipeline costs,
while others had traditionally segregated the costs of new pipeline projects, specifying various rate levels depending on the facilities needed to
serve customers that contracted with the pipeline at different times.[83] In
litigated cases, the FERC had displayed no particular consistency on the
matter.[84] But in the era before open access and the transfer of capacity
rights to contract shippers, the issue was not seen as a major problem.

The FERC's actions in 1985 and 1992 creating well-defined transport contracts for firm shippers greatly heightened the importance of
the issue. Shippers wanted their newly won capacity entitlements to retain their value, while pipelines wanted the value of that capacity to help
to underwrite the sale of new capacity entitlements at less than the cost
of construction. The nature of the dispute should have been elementary
from an economic perspective, as the practice of rolled-in pricing was
manifestly a potent and unjustifiable barrier to pipeline entry. Rolled-in
pricing would obviously damage the market for new capacity by heavily
favoring the projects of the incumbent pipeline companies with the low-
est-historic-cost existing capacity, whatever the actual incremental cost
of constructing the new capacity.

The battle between pipeline companies and their contract shippers

lasted from 1992 until 2000. In 1995, the FERC tried but failed to settle the issue in a policy statement by specifying that if the cost to existing shippers from a pipeline expansion project was less than 5 percent, then the FERC would make a presumption in favor of roll-in.[85] The
5 percent threshold altered behavior in the predictable way: for the next
few years, most pipeline capacity additions were targeted to affect the
cost of existing shippers by just less than the 5 percent threshold. The incumbent pipelines' victory on rolled-in pricing was only temporary, however. Various cases demonstrated to the FERC that any roll-in threshold
would affect pipeline behavior and damage the market for new pipeline
construction. In 2000, the FERC reversed itself, requiring that pipeline
companies segregate the new construction costs for the purpose of calculating distinct regulated charges for the new capacity.[86] By imposing
incremental pricing, the FERC allowed prospective shippers to decide
whether an incremental project is financially viable on its own merits.
For existing holders of firm entitlements, the new policy meant that the
value of the transport entitlements in the market would not be drawn
away by the pipeline companies to underwrite a barrier to competitive
entry in the construction of new transport capacity from one point to another. The value of those entitlements in the market for transport would
stay put with the holders of those entitlements—to use or trade as they
assess the value of those entitlements in the market.

Inventing a Fully Informed and Costless Trading System

The FERC's 2000 action required the expansion and modification of
consolidated pipeline reporting requirements to improve price transparency and allow more effective monitoring of possible discrimination and
the exercise of market power. The first issue involved confidential pipeline information (the FERC held that by definition there would not be
any). The second issue involved the trading platform.

A necessary element in establishing the market for the legal rights to
capacity is the free and transparent flow of information. The FERC confirmed its intention to continue to promote the most open and timely
electronic information system possible, confident that its vision for a
market in point-to-point transport capacity rights depended on it.[87] And
while the FERC acknowledged that some shippers thought that its infor-
mation-reporting requirement might cause some burdens, and also that
it could give shippers knowledge of their competitors' general market-

ing strategy, it was more than convinced by the need for the market to be fully informed to operate efficiently and to uncover undue discrimination or market manipulation when it appeared. Thus, the FERC chose to require the most comprehensive and immediate provision of all information on the identities and quantities, locations, and so forth, of all shippers. For the FERC, there are no trade secrets with respect to the use of the regulated interstate pipeline system—it is an open book.

In addition to making federally regulated pipelines open books for buyers and sellers of capacity, the FERC also required that the pipelines create web-based trading platforms (electronic bulletin boards).[88] Those bulletin boards have become the information and trading platform for the daily purchase and sale of transport entitlements on the regulated interstate gas pipelines.

The FERC Adjusts to Its New Role in Overseeing the Market for Entitlements

Setting incremental pricing as the default rule for new pipeline construction put an end to the traditional fights over licensing. If a pipeline company approached the FERC with a new proposal, it was no longer required to demonstrate connected gas supplies as it had once done before the era of open access. With a portfolio of letters of intent from committed incremental shippers, the FERC had prima facie evidence of economic need, which the Natural Gas Act still required that it establish. With incremental pricing and without the ability to draw upon the value of existing contracts to subsidize new construction, only the pipeline and those that committed to sign contracts for new projects (the "principals" to the transaction) had a stake in the matter. Other pipeline customers stayed out of the proceedings as the new capacity project did not affect them. The traditional sources of the seemingly open-ended litigation inherent in traditional licensing fights before the FERC vanished as a result. With SFV and incremental pricing, the pipeline would receive (and entitlement holders would pay) its cost of service, whoever held the capacity rights and whatever quantities of gas actually moved through the pipeline.

Gone too were the almost endless litigated rate proceedings on interstate pipelines, where coalitions of large shippers fought with one another over the allocation of costs on an essentially pooled pipeline system for delivering gas. Given a thorough examination of the cost of service, the tariff structure for existing pipeline capacity generally followed traditional cost allocation methods, and pipeline customers find little reason to fight over cost allocation or tariff design. In licensing, and tariff design, pipeline cases before the FERC have become somewhat perfunctory affairs.

This is not to say that the FERC has had little to do in overseeing this new market in legal entitlements to gas pipeline transport. It has been watching carefully the development of the market in transport entitlements, having to judge whether to cap the prices for capacity entitlements or let them trade at what the market would bear. Dealing with the issue in 1992, 2000, and 2008, the FERC had decided first to cap, then to deregulate for a temporary period, and finally to deregulate permanently prices in that market. In essence, the FERC has decided, with much experience to back up its deliberations, that the transport entitlement market needs no price or aggregation restrictions at all.[89] The FERC has also had to deal with the consequences of open access on pipeline customers' decision not to use certain links that the pipelines had traditionally used on their behalf. In essence, the ability of shippers—mostly gas distributors—to select their best routes created a shakeout. Some entitlements were worth less to shippers than their underlying cost and were "turned back" to their pipeline owners (particularly in the mid-1990s). The FERC had to deal with pipelines fairly while ensuring that the costs for those entitlements were not merely transferred mechanistically to the pipeline customers that remained on those unpopular links.

The FERC has also had to be on the lookout for any source of market abuse that would impair the functioning of the market for legal entitlements. That is, its new tasks have been related less to traditional rate regulation or licensing (which were no longer particularly controversial) than to the safeguarding and efficient functioning of the market in entitlements.

The Market for Entitlements Itself Learns and Adapts

It was one thing for the FERC to create the market in contractual entitlements for transport; it was another for those who bought and sold such entitlements to learn how to use or trade them effectively. Since the creation of these transport contracts, three highly visible shocks to the transport market in the United States show how the prices for those rights respond.[90]

The first example of a stress in the transport entitlement market occurred when the heating season of 1995–96 began with below-normal
temperatures. This resulted in large natural gas storage withdrawals that
could not be readily replaced. When temperatures again dropped dra-
matically across the midwestern United States, there was not enough
available gas in storage to meet the rising demand. Accordingly, gas traders panicked, and the Chicago city-gate pricing point spiked greatly relative to the Henry Hub in Louisiana (reaching a differential of ten dollars
per thousand cubic feet when the normal differential was a few cents).
It was a learning experience for gas traders. The cold snap in 1997 was
much like the one in 1996, but the gas market and traders had learned
from the year before, and the relative price spike in Chicago was only
one-fifth as high.[91]

The second notable stress occurred during the highly publicized California energy crisis of 2000–2001. Supply constraints, among other
factors, resulted in widespread electricity shortages across the western United States. Accordingly, the price of natural gas spiked because
of the increased value of electricity generated in natural gas–burning
power plants. The price of gas at the California border spiked hugely in
late 2000 and early 2001, with unheard-of needle spikes of thirty dollars
and almost fifty dollars per thousand cubic feet at the California border vis-à-vis the Henry Hub.[92] This differential prompted the quick planning, licensing, and construction of a major new pipeline expansion from
the Rocky Mountains to California, finished in 2003, the incremental expansion of the Kern River Pipeline.

The third and most recent stress on the gas system occurred in the
summer of 2005, during hurricane season in the Gulf of Mexico. During this period of already tightening energy supplies, two hurricanes dis-
rupted a large portion of the US natural gas supply and production. In
addition to completely shutting down the Henry Hub for a day and week,
respectively, Hurricanes Katrina and Rita led to different and larger-
than-normal supply-demand imbalances across the country and thus
larger price spreads in transport entitlements. But the market cleared in
both cases, as with the others, and the pattern of entitlement values for
transport to and from the different parts of the market returned to normal by January 2006.[93]

These events illustrated how the flexible and well-informed market
for contract entitlements learned to react to significant shocks in the
market for gas. In each of the three cases, the market responded to an

exogenous shock (winter peak, seasonal imbalance, or natural disaster) with the spot price for contract entitlements moving according to the local supply and demand for natural gas, and the free trade in available pipeline transport entitlements.

US Gas Distributors Make Highly Effective Pressure Groups

Effective pressure groups drove pipeline regulatory developments from the late 1950s onward to 2000. Unlike the oil pipelines, the US gas pipeline industry inherited long-standing and ready-made adversarial pressure groups of state-regulated gas distributors. These distribution companies were decades old when natural gas displaced their locally produced coal gas. For a while, it looked as if these distribution companies would become permanent parts of multistate, vertically integrated utility holding companies. But the Holding Company Act of 1935 returned those distributors to their independent, state-regulated status. These distributors, along with the consuming states and cities of the northern parts of the country, constituted the highly effective pressure groups that fueled the regulatory and legislative conflict from the early 1950s until 2000, after which the groups generally disbanded with the attainment of the market in legal entitlements for gas transport.

Location along the routes of the key interstate pipelines formed the basis for these distributor pressure groups. For example, the New England Customer Group was made up of the sixteen independent gas distributors that took gas in "Zone 6" on the Tennessee Gas Pipeline (which covered the New England states served by Tennessee). That group banded together to hire economists and litigate as a single party in cases before the FERC. The Midwestern Customer Group, the New York Customer Group, the Algonquin Customer Group, and others were all collections of distributors that funded independent counsel and experts to present their cases before the FERC in all proceedings involving gas pipelines and the move toward competitive gas transport.[94]

In the way these groups pressed issues before the FPC and later the FERC, they emulated the Associated Gas Distributors (AGD). The AGD—essentially a group of distributors in the northeast United States—formed in 1955.[95] It was to be highly effective, with Alfred Kahn as its witness, in opposing the deregulation of gas prices in the era before buyers and gas producers could bargain directly through open-access pipelines without the pipeline companies themselves acting as intermediaries. Jules Joskow, who in 1955 was on the economics faculty of New York University, described in his 1992 memoir how the AGD sought his help to counter the sophisticated presentations they expected on the question of wellhead price controls by the oil company experts.[96] Joskow admits that he and his associate Irwin Stelzer did not themselves know much about the oil industry. But Stelzer's thesis adviser at Cornell, Alfred Kahn, and his friend and collaborator, Joel Dirlam, knew a good deal. This group was then hired by those distributors. It was the AGD that funded Kahn's analysis and ultimately effective testimony in the various cases before the FPC dealing with the regulated wellhead price of gas. In their resulting investigations and presentations on behalf of that group of gas distributors, Joskow did not know that he and his colleagues were playing a part long held by Commons as a key role for economists in advocating for fair results in the economy.[97] In this respect, Commons would have seen the AGD's activities, with Kahn as the highly effective public witness, as part of the relationship between experts and pressure groups that he found so important in pushing for the orderly and efficient operation of markets.

Subsequent action on the part of those distributors was also important to erasing the ability of the US interstate gas pipelines to exercise market power. After the first general acceptance of open access by pipeline companies in 1986, it was the various groups of distributors that pressed the issues that led to the FERC's effective imposition of the commodities clause in 1992 (through "pooling points") and incremental pricing for capacity additions in 2000. As groups, these distributors and their experts, in operations and accounting, were also instrumental contributors in the long investigatory proceeding that settled the terms and conditions of the trade in entitlements (balancing, penalties, segmenting, notice, etc.).

Any economic history of the development of a market in legal entitlements to gas transport in the United States would be incomplete without recognizing the sustained collective action on the part of gas distributors and their state and municipal allies who acted in the interest of their constituencies of many millions of local gas consumers. It is no overstatement to say that the creation of the competitive gas transport market in the United States owes its existence to these doughty gas distributors, acting over decades through adversarial litigation—first in a contest against gas producers and then against the pipeline companies. They pushed for decades to erase the sources of market power or the barriers to entry that would keep delivered gas prices up and transport options restricted. The dearth of such sources of sustained collective action in other large-scale gas markets—principally in the European Union—may be the most insurmountable barrier to achieving such competitive pipeline transport markets.

The Unscripted Evolution of Competitive Pipeline Transport

The transition from the unregulated, vertically integrated pipelines before 1935—which Troxel, who had witnessed them, called irresponsible and notably acquisitive—to the competitive market in transport entitlements after 2000 was unscripted. No economist or legislator in the 1930s had any idea that the remedy for those abusive practices lay neither in common carriage nor in utility regulation, but in a market covering a bundle of intangible property rights that economists had not yet conceived.

Even without a script, however, an analysis of the entirety of the story leading from vertical integration to Coasian bargaining in gas transport seems to contain elements of inevitability. Congress in the 1930s had no real choice but to cleave pipelines from gas distributors, even if it would involve the most drastic state intervention into private business affairs that the nation had yet seen. Rejecting common carriage was also a fait accompli, not just because it had failed as a way to regulate oil pipelines but also because the nation's gas consumers (and their political and utility representatives) would never have accepted the risk of less than privileged access to the pipelines whose construction their credit had underwritten and upon which their constituencies would absolutely depend for their fuel.

But Congress's choice to regulate gas pipelines as utilities—the only other regulatory model around—was itself doomed to fail, as the semi-rival pipelines bought gas in a race to win new licenses, invest in new pipelines, and grow with a nation hungry for gas. That race skewed gas markets in the fields, and there was never any chance that the ultimate engines for pipeline industry creditworthiness—the millions of consumers connected to gas distributors—would agree to turn pipelines loose to buy gas as those pipelines wished. Regulating gas prices was itself no viable, long-term solution, for as reliable as the regulator was for the purpose of facilitating pipeline investment and restraining pipeline market power over captive customers, it was to prove no good at setting the price of gas.

Costly shortage or costly surplus was bound to result until pipelines were removed from the gas market to concentrate on transport. Accepting that change, however, invited more regulatory and legal conflict. The conflict ended when pipelines became merely owners and operators of cost-based, point-to-point regulated transport capacity, where shippers bought and sold effectively perpetual transport entitlements based on the value of gas in the nation's varied locations. The end of the conflict between shippers and pipeline owners also signaled a transformation of the regulator's prime job. That job had once consisted mainly of regulating entry and price, although much of the regulator's traditional work in the actual prosecution of regulatory cases was comparatively passive— watching from the sidelines as the principals (pipelines and distributors) battled. Now the FERC has another task—preserving the value of tradable entitlements for those who hold them.

Given the private and diverse ownership of the pipelines, the inherent power of the distributors (and consuming states) to shape public policy, and the role of the Supreme Court in defending regulated property from seizure, it is hard to see that any other transacting outcome would have resolved the high-stakes legal and regulatory conflicts that involved petroleum-producing companies, pipeline investors, and gas distributors. In 2000, the year that the FERC resolved the last outstanding issues and the competitive pipeline transport market took off, Oliver Williamson wrote that the movement that fueled the property rights literature among economists had "overplayed its hand" in implying that making markets by defining and enforcing property rights would be easy.[98] Williamson was right. True, the application of the theory has been highly successful in the vast US gas pipeline transport system. But from the first congressional debate to its realization, it took almost a century to accomplish.

The Competitive Potential for the World's Pipeline Systems

C ompetitive trading in bulk commodities on a continental scale needs competitive inland transport. The rise of vigorous gas trading on the New York Mercantile Exchange (NYMEX) accompanied and depended upon the rise of Coasian bargaining and transport competition. As such, the vigorous gas market, centered around the Henry Hub, merely confirms for gas what is true for other commodities that employ their own physical trading locations. Why haven't more jurisdictions investigated or capitalized on competitive Coasian bargaining in point-to-point pipeline transport to support competitive commodity trading? What nation's consumers would not want the competitive fuel commodity market and security of supply that such transport would support?

One answer must be that the US gas pipeline regulations seem impossible to emulate. The institutional origins and various conflicts surrounding the century of pipeline regulation in the United States could challenge anyone's understanding, let alone emulation. If that were not the case, someone would have called for the Federal Energy Regulatory Commission (FERC) or Congress to try to re-create its *gas* transport arrangements for its *oil* pipelines—and evidently nobody has. Another answer is that those who own or regulate oil and gas pipelines without such a market in legal entitlements may have no intrinsic interest in following that US gas pipeline industry example. Indeed, where that example may be understood elsewhere, it is also true that the elements of US gas pipeline regulation constitute a threat to the value of invested capital. A working market in point-to-point transport entitlements erases pipeline market power and the regulatory status quo—the comfortable and established ways that governments regulate their pipeline businesses.

This chapter examines whether other pipeline systems would seem to have the potential to support competitive pipeline transport. To be sure, the Coasian market appears to be an optimum solution to the market for the long-distance pipeline transport of fuel. Comparing any other existing pipeline transport system to that Coasian example spotlights the entrenched interests that constitute barriers to a more efficient outcome. The following sections, which describe some of these entrenched interests, may seem to suggest that there is little room for more efficient pipeline transport in those other pipeline sectors. Then again, a functionally competitive pipeline transport market would have appeared to face impossible barriers in the United States in 1960. Effective collective action and political will, in the right circumstances, can topple such barriers.

Oil Pipelines in the United States: A Century of Evolving around Common Carriage

The issue of oil pipeline market power has never gone away in the United States. It continues to be rooted in the Hepburn Amendment's common carriage restrictions on reasonable pipeline contracting and the inevitable vertical integration that results. Congress, the Department of Justice (DOJ), and the FERC have essentially accepted that a century of industrial ossification is far too much to undo, even if the problems involved are well understood. Instead, all three bodies have decided to examine and deal with the symptoms of market power rather than its roots.

What if Congress could snap its fingers and re-create the gas industry system of tradable transport entitlements for the oil transport system? The FERC in any event would face large transition costs in any attempt to impose Coasian bargaining for capacity for the US oil pipelines after a century of industry development under common carriage. John R. Commons said long ago that private property gave stability to economic relations.[1] Private property has been very clearly defined by the courts in the United States over the course of a century of utility regulation. The FERC cannot change its style of regulation on its own—it acts at the direction of Congress. And Congress itself is generally barred under the

US Constitution from changing a style of regulation that would damage the value of the property of those it regulates.[2] It is almost inconceivable that the vertically integrated oil pipelines in the United States would acquiesce to a new regulatory regime that would make them builders and operators of a transport system that creates tradable property rights (as the independent gas pipelines did only under duress). Even if Congress could resist the oil industry pressure groups (which it has often not done in the past), a revocation of the Hepburn Amendment, and the widening of the Natural Gas Act to include oil pipelines, would create many years of constitutional litigation. Even if Congress succeeded in the end, the industry would have to face a market that might not agree with the industry's historical decisions on where to locate pipelines and how big to make them. That resulting shakeout over "stranded" assets would involve more litigation still.

It is a serious question whether the imposition of Coasian bargaining would be worth the cost of upending the industry's distinctive and long-standing institutions. The DOJ has been no friend of the integrated oil pipelines, and yet it found in 1986 substantial competitive potential in the industry. In addition, the FERC's regulation of oil pipelines since 1978 has partly sorted out the regulatory mess it inherited from the ICC, including the settling of the value of the property inherent in each oil pipeline's tariff base. The FERC has been willing to somewhat lessen the risk of financing new common carrier lines through the advance approval of key rates and tariff issues before construction.[3] Further, the comprehensive 2008 FERC action regarding the pipeline with the biggest tariff base of all, the Trans Alaska Pipeline, seems to have settled issues of reasonable capital structures and rates of return that have beset and confused the industry since the 1941 consent decree.[4]

Without Coasian bargaining in oil pipeline transport, the sort of traditional regulatory fights over entry and cost allocation that arose in the pre-open-access gas pipelines will continue to arise for oil pipelines.[5] But substantial competition that already exists for oil pipelines, which includes other modes of shipping oil (such as barge, tanker, and truck), and the sharpening of regulatory methods should reasonably restrain oil pipeline market power, even if that restraint is accompanied by the rather ponderous FERC regulation of per-barrel tariffs. Common carriage remains a poor way to regulate major pipelines, impairing as it does the long-term contractual relationships that would allow one pipeline owner to serve independent and competitive shippers. It would doubtless be an improvement in the competitive use and expansion of crude oil and products pipelines if Senator Lodge's 1906 Hepburn Amendment were wiped away in favor of a move toward a market in such transport entitlements. Had the oil pipelines come under utility-style private carriage regulation in the 1930s, as the gas industry did (say, as a result of Thurman Arnold's efforts that were sidetracked in 1941), the level of vertical integration in the US petroleum industry generally would almost surely be smaller. But such a radical regulatory change in the twenty-first century for an industry that is in some cases reasonably responsive to fuel markets with century-old industry structures and institutions is unlikely.

Gas Pipelines in Canada: Open Access without a
Market in Transport Entitlements

Canada has an extensive interprovincial gas pipeline system dominated by the TransCanada pipeline. TransCanada traditionally shared the private carriage role of its US counterparts—buying gas in the producing fields and selling to gas distributors and others. Prior to 1985, the regulated commodity price of natural gas in Canada was set based on crude oil prices by agreements between the federal government and the province of Alberta, where most of the gas originated. A number of political and economic exigencies led to the formulation in October 1985 of what has been known as the "Halloween Agreement" (i.e., the Western Accord on Energy Pricing and Taxation), signed by the federal government and those of the three major gas-producing provinces: Alberta, British Columbia, and Saskatchewan. Most notably, the Halloween Agreement deregulated the price of gas.[6] Subsequent Canadian regulatory reform unbundled gas sales and transport services previously sold as a single product by merchant pipelines.[7]

A resale market for the transport rights began in 1989, when the National Energy Board (NEB) approved changes on the TransCanada system to allow shippers to reassign firm transportation rights to third parties.[8] The NEB allows capacity release but does not require Canadian gas pipelines to assign this right to shippers. The NEB decided in 1995 that the posting of capacity trades on electronic bulletin boards was unnecessary and should not be required of pipeline companies.[9]

From a constitutional foundation through to administrative practices, accounting practices, and judicial review, Canada and the United States have virtually indistinguishable regulatory environments—so much so that the *Hope* decision is cited in Canadian rate cases.[10] The regulatory environment in both countries is shaped by judicial decisions and includes the right to earn a "fair return" on investment, as determined by the opportunity cost of capital, which in Canada is termed the "comparable investment" standard. Indeed, the Canadian decisions are widely recognized as establishing an effective protection for the rights of regulated property owners that is very nearly identical to that of the United States.[11]

The similarities in the regulatory environment between Canada and the United States generally stop when the issue turns to the transparency of the transport system and the resale of capacity rights. The NEB itself noted in 1996 the scarcity of reliable information on the size and liquidity of the gas pipeline capacity resale market, which it believed was quite active. The NEB backed up this statement by pointing out that as of fall 1995, there were four hundred TransCanada firm service agreements outstanding, of which more than one-third were under temporary assignment, to TransCanada's knowledge.[12] However, many of the contracts for capacity release were created through verbal agreements. As such, and as the NEB pointed out, information on this market was not reliable.

Four of the elements of the market for legal transport entitlements in the United States—utter transparency regarding available capacity and the price of trades, a frictionless web-based exchange, an effective commodities clause, and incremental pricing of new capacity—do not exist in Canada.[13] Without those elements of a system of property rights, there is no definitive split between pipeline ownership and the ownership of legal transport entitlements. The property rights inherent in shippers' transport rights would not effectively trade in the United States without those elements, and it is not surprising that a similar market has not developed in Canada, even though there is a significant amount of capacity trading that happens with the cooperation of the gas pipelines involved.

Since Canada is an integral supplier to a huge gas market to the south, which does display a vigorous market in legal transport entitlements, the absence of Coasian bargaining in Canadian pipeline transport rights probably has little effect on the gas prices at various locations across Canada. Gas prices at locations just inside the border (where the great concentrations of Canadian gas consumers are located) do not stray in significant degree from competitive gas prices in the United States. As a result, there is little pressure to alter the open-access gas transport arrangement that already exists in Canada. There is no evident movement toward the type of split between pipeline ownership and the ownership of transport entitlements that developed by 2000 in the United States. The lack of such a market, however, may become an issue if a gas pipeline is ever built from Alaska through Canada to the lower forty-eight states.[14]

UK Gas Pipelines: Abstracting from Transport

It is easy to criticize the UK government in hindsight for creating a privatized British Gas without giving serious consideration to any realistic open-access obligation for independent gas sellers.[15] It is harder to appreciate the choices available to the UK government from its 1980s perspective (even in 1986, for example, the United States had not yet succeeded in implementing open access itself). Nevertheless, the privatization was a major missed opportunity to create a more competitive and efficient industry structure.

The privatized British Gas was a vertically integrated gas supply, storage, transport, and distribution company. The privatization produced no independent distributors that could play the role of pressure groups on behalf of connected customers. The UK government established no accounting or regulatory institutions upon privatization that would create unambiguous property values for the capital invested in pipelines and hence facilitate arm's-length transacting by contract. As a result, the UK government and its regulators focused instead on lowering the entry barriers to gas marketers even if they had to mask the considerable transport cost of doing so through the entry/exit tariff model.

Institutional Attachment to Abstracting from Transport

Since its first adoption of entry/exit, the pipeline charging regime for gas transport in the United Kingdom has increased greatly in complexity.[16] Regulated charges are now split between the pipeline owner and the pipeline "system operator," each with its own set of charges. The basic set of entry and exit tariffs comes from a model that purports to calculate the long-run marginal cost (LRMC) at each entry and exit point.[17] Entry capacity is sold through five related auction mechanisms, stretching from between one day and sixteen years, subject to "reserve prices" coming from the model. The costing, pricing, and capacity allocation system is highly complex, requiring ongoing seminars and training sessions for even sophisticated industry participants to understand. Besides effectively blockading independent pipeline entry, the complexity accomplishes little or nothing in terms of the efficient use and expansion of the gas transport system in the United Kingdom.

Why is the charging method for the UK gas transport system so complex? More than anything, it reflects the regulator's desire to pursue the neoclassical trappings of efficiency without considering how constraining its regulated tariff making is in practice. The need to tie all charges to the pipeline company's permissible revenue level makes most of the work of a gas pipeline LRMC model irrelevant (especially so, if it is unrelated to particular routes through the pipeline system). The capacity auction mechanisms are complicated and unsupportive of the type of long-term relationships with predictable charges between pipeline companies and major users that asset specificity would otherwise require.[18] Those in the United Kingdom who push such periodic auctions evidently believe that they are efficient methods of transacting. But Coase saw through such fallacies in 1937. He and subsequent economists who understand the cost of contracting would see that imposing such auctions on pipeline commercial relationships is inconsistent with the asset specificity inherent in the inland gas transport industry and thus counterproductive to an efficient utilization of those specialized transport assets.

Setting the Structural Stage for Regulatory Difficulties

The rush to privatize British Gas had two consequences that colored other subsequent international efforts to privatize pipelines. The first was the government's failure to restructure before privatization. The push to open the system to third-party gas shipments, at all costs, led to the creation of a tariff regime that ignored the transport system's basic operation. That tariff regime has proved impossible to uproot despite its manifest inability to transmit effective gas transport price signals or constitute an orderly and stable relationship between the gas pipeline company and its shippers. Also, by failing to impose structural separation of distributors from transport pipelines, as well as the "commodities clause"—which would have forbidden British Gas from owning the gas it shipped in its lines—the privatization inhibited the growth of any effective private pressure groups. This meant that any disputes over the soundness or great cost of the entry/exit system would take place only between British Gas (later National Grid Gas) and its regulator (first Ofgas, later Ofgem). Neither the pipeline owner nor the regulator has yet shown an interest in making gas transport in the United Kingdom more cost based, competitive, or efficient. Doing so would require abandoning an institution that both parties, perhaps not surprisingly, seem to like: countrywide entry/exit tariffs that effectively block competitive entry and make for much staff work.

The second consequence of the rapid privatization of British Gas was the adoption of an existing public-enterprise-oriented accounting study (known as the "Byatt report") that was inconsistent with defining regulated private property in a way conducive to orderly regulation or arm's-length transacting by contract with regulated enterprises.[19] Among the institutional foundations for such transacting, two are fundamental: transparent legislated accounting standards and a reliable (constitutional) definition of private property that avoids the uncertainty and/or circularity inherent in using deemed property values, rather than reasonable earnings, as the basis for regulating prices.[20] Neither at privatization, nor now, does the United Kingdom mandate accounting standards for its regulated businesses, including pipelines. Further, property values do not reflect the opportunity cost of invested capital in the United Kingdom, as the *Hope* decision requires in the United States, but rather depend on an independent valuation determined by Ofgem from tariff review to tariff review. From the perspective of the "end-results test" of the opportunity cost of capital on accurate and objective book equity that defines the *Hope* decision, the resulting pipeline valuations are vague and subjective.[21] Even if the United Kingdom scrapped entry/exit and auction-based third-party access (TPA) in favor of a point-to-point transport regime of long-term, arm's-length contracts, problems would remain because of weak institutions supporting regulated private property rights. The inability to establish unambiguous property values as the basis for contracting would impair both the predictable regulation of the existing pipeline system and the value and tradability of those contract rights. That is, without a clearer and more dependable constitutional definition of the value of regulated property in the United Kingdom, it would be very difficult to support the effective assemblage of a bundle of long-term transport rights that could give rise to successful Coasian bargaining and competitive transport.

The United Kingdom has multiple access points for gas to its island as well as direct pipeline connections to continental Europe and Ireland. Even with the island's limited geography, there appears no physical or technical reason to dismiss the potential for more competitive arm's-length transacting for the use and expansion of the pipeline transport system. But against the structural and institutional impediments created at the time of privatization, the prospect for defining inland gas transport in the United Kingdom that would promote such a goal is remote.

Overcoming Structural and Institutional Barriers in Australia

Eastern Australia has an extensive pipeline system and growing use of gas, but it has little competitiveness in either gas supply or gas transport. Since the Hilmer report, the major gas-consuming cites have gone from being served by government-owned gas pipeline monopolies to being served by two private, generally unregulated gas transport pipelines from a highly concentrated, joint-ventured gas-producing sector. It is hard to facilitate competition in gas pipeline transport with so much concentration in gas supply. But it is also hard to foresee any particular movement in promoting competition in the gas-producing sector when the highly concentrated pipeline sector remains inscrutable and in control of the transport rights on its pipelines.

This is not to say that pipeline tariff regulation had any greater chance of forming the basis for Coasian bargaining than is evident in the United Kingdom. The accounting basis for regulating transport tariffs in Australia is ultimately rooted in the valuation philosophy for public enterprises espoused prior to privatization in the United Kingdom as described in the Byatt report. That report was subsequently adopted in Australia after the application of "optimised deprival value" tariff base accounting (a Byatt report concept) in New Zealand. For example, in a September 2003 tariff case, through vagaries in the accounting rules for valuing the Moomba-Sydney pipeline tariff base (again, based as in the United Kingdom not on reasonable earnings on invested capital but on deemed regulated property values), the owners of that pipeline requested A\$834 million, whereas the Australian Competition and Consumer Commission (ACCC), the federal regulator, deemed the assets to be worth only A\$545 million. The company appealed to the Australian Competition Tribunal (ACT), which adopted the higher figure.

The ACCC appealed to the next court, the Full Federal Court, which in 2006 reinstated the lower figure. The company appealed further, and the High Court of Australia reinstated the higher figure.[22] Such a game with A$289 million in one pipeline's tariff base, in a single case, is a good example of the type of quixotic uncertainty in valuing regulated property that Bonbright (in his 1937 *Valuation of Property*) and later the US Supreme Court (in its 1944 *Hope* decision) sought to avoid.

The lack of a competitive gas sector in Australia is manifested in the lack of a market-based gas price-setting mechanism. The gas prices in every region of the country are today set in periodic complex and lengthy high-level arbitrations. Given the number of producers and existing pipeline links between gas fields and consuming areas, Australia would appear to have the structural basis for a competitive gas market. The barriers to such a market in Australia are institutional. The ability for Australia to exploit any competitive potential in its gas supply will probably remain unfulfilled as long as (1) two producing consortiums dominate most of the flowing gas in eastern Australian markets (although there is some hope for independently produced coal seam methane) and (2) the pipeline system remains relatively opaque and unregulated—or without the ability to offer contracts in the case of Victoria—preventing gas distributors, power plants, and others from contracting for long-term transport, at transparent and predictable regulated prices.

Argentine Gas Pipelines: Institutional Difficulties and Expropriation

Argentina has a venerable gas industry and some of the world's longest and oldest gas pipelines. Furthermore, at the time of privatization, the Ministry of Economy and its divisions (like the Secretariat of Fuels) were dominated by a number of well-trained economists, in high government positions, who looked to a respected and senior professor of economics at the University of Córdoba, Carlos Givogre, for advice. Those economists ignored the representations of some vertically integrated European gas companies and restructured the state company prior to privatization—specifically to promote pipeline rivalry, impose the "commodities clause," and set up independent distributors as the main bargaining counterpart to independent producers.[23]

But while the government of Argentina could achieve success in establishing the structural foundation for an efficient and competitive gas pipeline sector, it could not avoid trouble with maintaining the meticulous institutional foundation for regulating investor-owned enterprises.[24] A key example lies in accounting. The initial privatization was accompanied by sets of accounting books that would have been familiar to US regulated pipelines.[25] But the purchasers of the privatized transport and distribution companies were mostly European gas companies. With such owners, and a largely inexperienced regulatory staff, interest in maintaining those regulatory and tariff accounts waned on both sides. When the regulatory agency itself failed to meet its own deadline to produce a definitive decision on the maintenance of regulatory accounts, continuity in the maintenance of transparent regulatory accounts for tariff-making purposes was largely lost.[26]

Ultimately, more damaging to the success of privatization in Argentina was the government's inability to maintain a stable macroeconomic and currency system. It was the problem with the currency that ultimately destroyed the creditworthiness of the privatized gas businesses in Argentina and their ability to grow and serve the public during the first years of the twenty-first century. Prior to the privatization of Gas del Estado, Argentina experienced what is perhaps history's worst period of sustained hyperinflation, from 1970 through 1990, when the country's twelve-month inflation rate hit its peak of 20,266 percent.[27] In cumulative terms, it would take the equivalent of a stupefying 172 billion old pesos in 1991 to equal the purchasing power of one in January 1970. Argentina dealt with its inflation problem by pegging the Argentine peso to the US dollar and imposing strict institutional constraints on exchange rates, as well as banning peso-inflation contract provisions, as part of its 1991 Convertibility Law. It was clear to both international investors and Argentina's advisers (from the World Bank and elsewhere) that successful privatization required prospective utility investor-owners to defend against such currency instability. The US dollar calculation of tariffs in all of Argentina's utility privatization concession contracts (with conversions to pesos only at the time of billing) seemed to deal with these concerns. Throughout the 1990s and well into the year 2001, Argentina maintained its peg to the US dollar, despite growing evidence that the resulting overvalued peso was pushing the economy deeper into recession.[28]

By December 2001, Argentina's economy was in crisis, and in early 2002 it collapsed. Under the Public Emergency Law, passed on January 6, 2002, Argentina unilaterally suspended the US dollar terms in utility concession contracts and called for a renegotiation of terms in all privatized concession contracts. The effect of de-dollarization was immediate: with a 70 percent devaluation of the peso, the move deprived the utilities of most of their dollar cash flow and led to rapid default on their dollar obligations. The effect of the call for renegotiated concession contracts was slower, although a number of settlements on new concession contracts had been completed by the end of 2007, by which time the share values in Argentine utility equities had generally rebounded to the levels above those in late 2001, before the Emergency Law.

As a result of the Emergency Law, the Argentine government attracted a large number of treaty claims for expropriation. As of November 2004, seventy-four cases were pending before the World Bank's International Centre for Settlement of Investment Disputes (ICSID), of which thirty involved claims against the Argentine government by oil, gas, and utility companies looking to be compensated for losses incurred subsequent to the Emergency Law.[29] A number of decisions in those cases awarded many hundreds of millions of dollars in damages to the owners of the privatized gas pipeline and distribution companies.

The decline in the prices of utility shares leading up to the currency crisis in late 2001, before the Emergency Law, demonstrated that concession contracts notwithstanding, the fortunes of regulated utilities rose and fell with the economy. While the Emergency Law brought this home in stark fashion in early 2002, it merely confirmed the inevitable tie between a fragile economy and prospects for the mostly foreign investor-owned utility businesses in it. When an economy is in crisis, and public opinion is sharply against transfers from consumers to investor-owners of transport infrastructure assets, the public comes first. That was as true when Argentina failed to live up to its concession contract obligations in 2002 as when New York failed to live up to its Erie Canal bond obligations in the depression of 1839–42.

Argentina's deep institutional failure wrecked the tenuous credit that the regulated gas pipelines had maintained. Extraordinary efforts will be required to attract the kind of capital that Argentina needs to maintain and expand the pipeline system. The right industry structure is in place, and the regulator has had time to mature and develop. But the government of Argentina's credibility in protecting foreign investments was destroyed twice in twenty-five years, wounds that will stunt the efficient and economical development of Argentina's pipeline system for the foreseeable future.

Gas Pipelines as Barriers to a Competitive European Gas Market

If there is a part of the world that could benefit from creating the conditions for competition in gas supply among diverse arrays of suppliers, it is continental Europe. The EU has evolved institutionally to pool the sovereignty of its member states in a large number of policy areas, but its original and continuing focus is to create a common market to erase the internal barriers to trade in products that its businesses and citizens produce and consume. For a variety of reasons, its progress in energy markets has lagged behind that in others. However, the countries of the EU have internal sources of gas as well as four major external gas suppliers—none of which account for much more than a quarter of the EU's gas.[30] There are also prospects for major sources of gas entering the EU through Turkey from the Caucasus and the Middle East. In addition to pipeline supplies, the EU has many sites to store gas underground near the market areas and more than a dozen working LNG terminals that supply 10–12 percent of current EU consumption. The continent almost certainly has the supply diversity and the pipeline hardware necessary to create a competitive gas market. But the EU has yet to achieve any noteworthy movement in that direction. As a result, EU authorities remain deeply concerned about not only the basic lack of competitiveness in gas supply and transport but also the fundamental security—political and physical—of the EU's gas supplies, particularly with respect to Russia, its largest single supplier.

The extent of vertical integration across the continent is evidence of the lack of pipeline competition in the EU. Many gas distributors are part of large-scale pipeline suppliers (as in France and the United Kingdom). Europe does indeed have a large number of municipally owned distribution companies (as in Germany and the Netherlands).[31] However, these locally managed public undertakings have generally been stripped of their ability to contract for gas or pipeline transport capacity—and therefore the ability to act as organized agents or advocates for their connected consumers in the manner of distributors in the United States and Canada—as part of the efforts in the various legislative packages for "full retail access." The major "interconnectors" (that is, the pipelines not subject to third-party access rules) are owned in whole or part by the major producing interests or their affiliates in combination with the vertically integrated national gas companies. Most troubling of all is the extent to which the governments of the EU member states have allowed Russia's Gazprom to integrate downstream into their pipeline companies and vertically integrated gas suppliers.[32] It should be worrisome that a major continental gas market area like the EU would permit the gas supplier representing what it considers to be the principal threat to its supply security to obtain ownership interests in so many of its pipelines and pipeline affiliates.[33]

What if the EU could snap its fingers and split pipeline ownership from the ownership of legal transport entitlements? Such a move would permit its four principal suppliers (Norway, Algeria, the Netherlands, and Russia), as well as the smaller internal EU suppliers, to compete with one another, with respect both to flowing gas and to the prospect of making new pipeline links to other regions, like Turkey and the Caucasus. The trade in those legal entitlements would reliably signal where new capacity entitlements were worth the construction cost. There is probably sufficient supply diversity to allow EU competition authorities to be reasonably confident, given sufficient vigilance, about the potential for gas supply competition if the pipeline system were itself competitive and transparent. By creating a transparent system of EU-wide legal transport entitlements, market power for both pipelines and producers, and the security-of-supply worries about the latter, could evaporate as they have in the United States. Furthermore, as in the United States, transparent transacting on the EU's pipelines would make whatever market abuses do arise easier for the EU's competition authorities to discover. There appears to be no physical, operational, or technical reason why such a scenario is impossible.

The contrast with the United States is greatest in the institutions surrounding the EU's pipelines. In Europe, all of the major gas transport pipelines were built by governments or state-owned companies.[34] The EU has no strong authority to prevent member states from erecting barriers to interstate trade. Indeed, the EU has no strong continental regulatory authority at all—pipelines respond to a complex patchwork of EU and national regulations.[35] Transparency in transport capacities, costing, and pricing that are expected in the United States are almost totally lacking in Europe.[36] There are no effective pressure groups representing consumers. Europe has nothing even remotely similar to a system of tradable legal entitlements for gas transport. In the EU, some short-term swapping of pipeline capacity does occur, but it is effectively at the discretion of the pipelines. The impediments to creating such a market in the EU are institutional, not physical.

Inherent Institutional Barriers to a Market in Transport Entitlements

The market envisaged by Coase requires careful physical and legal definition, for ambiguity in either respect inhibits trade. From 1985 to 2000, great effort in the United States was devoted to refining both the physical and legal definitions of those entitlements. If the EU wished to promote the development of such a market in the rights to inland transport, what obstacles would lie in the way? Competitive trade in inland pipeline transport in the EU, and hence in gas, faces six institutional obstacles.

SPLIT JURISDICTION.   Perhaps the most daunting barrier to creating a competitive market in transport is political—the split in jurisdiction between the EU and its member states regarding pipeline regulation. The EU treaty has nothing like the commerce clause of the US Constitution, which gives the federal government sole and unambiguous jurisdiction over all interstate trade. The EU competition authority knows that this is a potent problem.[37] Chapter VIII of the third legislative package of common rules for gas regulation, approved in 2009, asks member state regulatory authorities to share similar policy objectives, perform a long list of common duties, and "closely consult and cooperate with each other . . . with any information necessary for the fulfillment of their tasks."[38] But such provisions are inherently weak in the face of the evident desire of national governments and national regulators to protect their "national champion" vertically integrated pipeline companies. Regarding uniform regulation across the member states, the third legislative package is little more than a general exhortation to the national regulators to do their best. There are no particular consequences if they fail or if they show preference to procedures that benefit their own domestic firms. The power to regulate pipeline prices and access terms to the EU's gas pipeline system remains within the national regulatory authorities, however they may choose to do it.

THIRD-PARTY ACCESS (I.E., COMMON CARRIAGE).[39]   The third legislative package requires regulated third-party access (TPA) for gas pipelines in the EU.[40] It also states, "Further measures should be taken in order to ensure transparent and non discriminatory tariffs for access to transportation. Those tariffs should be applicable to all users on a non discriminatory basis."[41] A market in legal transport entitlements requires something quite different: that transport entitlements rise to the level of tradable property rights, which the enforcement of the directive's common carriage–like provisions would prevent. As this volume illustrates, the need to finance pipeline capacity under such transacting restrictions leads to an inexorable pull toward either vertical integration or the elimination of pipeline rivalry. In other words, the TPA rules that lie at the heart of the EU regulations bar the formation of the property rights in transport, and that barrier effectively bars the development of the kind of competitive gas transport that would support a vigorous gas commodity market.

VERTICALLY INTEGRATED GAS PIPELINES.   Europe never experienced the problems caused by privately owned, vertically integrated utility holding companies that beset the United States in the 1920s and 1930s, and thus its policies regarding the tie between pipeline companies and gas distributors are milder than those that developed in the United States. Nevertheless, the EU competition authority has been strongly critical of the "vertical foreclosure" inherent in the gas pipeline system.[42] The third legislative package recognizes that vertical integration is a problem, and the EU continues to press for "ownership unbundling" as a remedy, although that new proposal focuses more on the functional unbundling of gas pipelines from commodity suppliers than on the structural unbundling of pipeline companies from gas distributors.

In either case, the EU's legislation pressuring for ownership unbundling contains an annihilating loophole: it permits pipeline owners to submit their operations to an independent operator rather than to separate.[43] To the extent that the EU competition authority wished to see rivalry in pipeline transport and new entry, independent system operation will halt that initiative for all practical purposes. Such "independent" operators are common for electricity transmission grids, which is evidently where the idea arose.[44] Transacting for gas transport, however, is different than transacting for electricity grid transmission—physical paths are well known for the former but quite impossible to predict with today's technology for the latter. Independent system operators handle electricity grid transmission precisely because of the inability of electricity transmission network users to contract for particular facilities—a basic reason that does not apply to gas transport pipelines.

Owners of integrated pipelines can be expected to use the loophole not only to retain their vertical relationships but also effectively to avoid the pressure of pipeline entry. The hostility of independent system operators to pipeline transport rivalry surely has complex roots—interlocked with politics, organizational governance, and pressure groups. The underlying purpose of an independent system operator is to manage the operation of a monopoly. On a competitive pipeline transport system, where the ownership of transport entitlements is split from pipeline ownership and operation, any nonpipeline "system operator" is redundant. The United Kingdom and Victoria have independent gas pipeline system operators for which monopoly management is their main purpose— and whose organization and jobs depend on the maintenance of that monopoly. The opportunity to derogate from the ownership-unbundling rules, in favor of independent system operators, will be seen by integrated pipeline companies as a highly attractive option to avoid unbundling and a means to maintain the effective cartelization among pipelines and gas sellers that characterizes that market. The proliferation of system operators will inevitably dampen transport rivalry in the EU, even if it may have some effect on the grossest forms of favoritism of pipeline affiliates.

PROVISION OF INFORMATION ON PIPELINE TRANSPORT. The markets in legal entitlements envisaged in 1960 by Coase are entirely dependent on transparency and the free and immediate flow of information. At the FERC, there are no trade secrets with respect to the use of the regulated pipeline system—it is all an open book. The EU evidently does not agree with the need for such transparency. Neither the EU nor any of the member states have imposed a mandatory accounting system like the Uniform System of Accounts for pipeline companies. Further, only "competent authorities," and not the public, are authorized to obtain the information that pipeline companies may maintain. Finally, the EU warns any national regulatory body against requiring the provision of any "commercially sensitive" pipeline information.[45] Taken together, these provisions leave pipeline companies in virtually complete discretionary control of the kind of operating and financial information that would support pipeline rivalry.

The EU competition authority has written about its underlying frustration regarding the provision of information on the European gas system—describing the contrasting views between those who believe that a functioning gas market needs more transparency and those who believe that such transparency could support collusion.[46] Concern about the provision of public information on shipments possibly supporting collusive behavior is not limited to Europe.[47] Such information may or may not facilitate collusion. But if the payoff for collusive behavior is substantial, it may well happen anyway in private. The provision of such information to other parties that might be harmed remains one of the only effective ways for those harmed to discover and remedy competitive abuses. Apart from possible collusion, however, it remains true that a market in legal entitlements cannot function without the kind of transparency that the FERC has required and for which the EU has provided considerable loopholes. There has been quite a bit of administrative action in Europe after the adoption of the third legislative package in 2009 devoted to harmonizing the provision of regulatory information across the member states. Organizations such as the Council of European Energy Regulators (CEER), the Agency for the Cooperation of Energy Regulators (ACER), and the European Regulators' Group for Electricity and Gas (ERGEG) all have mandates to promote cooperation and information provision among national regulators. All of this seemingly busy activity notwithstanding, the control of both accounting and operational information under the third package remains with the pipeline companies.

FUNCTIONAL SEPARATION (THE "COMMODITIES CLAUSE").   As far as the various legislative packages and more recent EU publications are concerned, there is considerable discussion about fair access to pipeline systems—but almost no discussion of the competitive problems that arise when pipeline owners ship their own gas. The lack of functional separation creates problems, as the EU *Competition Report* has itself recognized when it found evidence that pipeline owners have granted substantial discounts to their own affiliated shippers.[48] A number of cases of discriminatory behavior are cited in that report, all stemming from the lack of structural separation or the lack of functional separation between transport and gas commodity sales. Within many vertically integrated gas companies, trading names, brands, and logos are still shared. There is no application of a "commodities clause," which would prevent transport pipelines from owning the gas shipped in their trunk pipelines and effectively end the opportunity to extend obvious or subtle preference to affiliated gas suppliers.

PIPELINE PROPERTY AND THE REGULATION/ADMINISTRATION OF RATES. In the EU, the administration of regulated prices is relatively new and inconsistent across the member states. Predictable licensing, accounting, and regulated rate administration do not exist for new pipelines. The third legislative package calls for "published tariffs, applicable to all eligible customers" (Article 32, Section 1) but does not further describe the rate-making formula or rules on the level of permissible revenues. There is nothing approaching the *Hope* decision in the EU that would unambiguously settle, across that common market, the question of the value of regulated property. It is reasonable to expect that jurisdictions that have difficulty defining property rights for regulated property would also have difficulty forming contracts that could deal with asset specificity.[49] Without a firm base for valuing administered regulated property, the prospect that contract-based access to the continent's pipelines would form the basis for Coasian bargaining is not bright.

Another potent barrier to competitive inland gas transport across the European Union lies in the apparent wholesale adoption of the UK-style entry/exit pricing in the 2009 legislative actions, to the specific exclusion of pricing based on the use of particular point-to-point pipeline facilities—the basis for such competition in the United States.[50] Some European economists recognize the limits that entry/exit pricing places on a pan-European gas-trading regime.[51] As in the United Kingdom, such an oddball pricing scheme for inland transport, comprising otherwise easily identified capital-intensive pipeline facilities, is more a barrier to pipeline entry than anything else—particularly in its manifest inability to reflect any sort of efficient pricing and utter hostility to forming reasonable long-term capacity contracts. But as of 2009, as is the case with the other legislative barriers to a workable pan-European gas market, the collective action of the protectionist interests quite evidently makes a more persuasive case in shaping EU policy.

The Prospect for Pipeline Rivalry—Competitive Transport—in the EU

It may be uncharitable to be so critical, in the space of just a few pages, of the considerable efforts of those in the EU who have designed the three legislative packages for gas pipeline regulation with such evident hopefulness. But from a wider perspective, even the updated and expanded third package will not have its stated effect of promoting either gas competition or greater supply security within the European common market. The EU authorities simply face too many institutional barriers to such competitiveness. What is more, to the extent that those in the EU who wish for gas competition continue to push for "a true European end-user market" policy, they will only further deprive themselves of possible distributor allies as a supportive pressure group in their efforts to pursue competition in either gas or pipeline transport.[52] Indeed, the principal innovations of the third package—loopholes in the strict vertical unbundling requirements and the blanket imposition of entry/exit pipeline pricing—are retrograde steps in the pursuit of either competition or security. System operators may prevent the most obvious affiliate favoritism, but if those operators emerge, they will constitute a new institutional barrier to competition in addition to their sizable bureaucratic cost. Those new independent system operators, cleaving to their entry/exit models, are apt to become as difficult to eradicate in any future move toward a more competitive transport system or gas market in Europe as any institutional barriers that already exist.[53]

As it stands, intense political maneuvering accompanies any plan to expand the EU's gas supply lines, particularly to the east. Gazprom wants to avoid the possible holdup from transit countries of Poland and the Ukraine by building an undersea supply line in the Baltic to Germany. The US State Department, among others, is trying to promote another gas supply line that bypasses Russia to connect southeastern Europe to the Caspian region and the Middle East through Turkey. That the US State Department is interested in gas pipelines thousands of miles away, for what is essentially a regionally produced and consumed commodity, is proof enough of the complicated politics of Europe's gas supply. Multiple and redundant pipelines, built on speculation and supported by sovereign guarantees rather than commercial contracts, may promote some sort of regional rivalry and supply security for Europe. But it is an expensive and uncertain solution, because it ties up so much capital without the commitments that asset specificity would otherwise

demand. The manifestly less costly option is to use the continent's existing pipelines more competitively.[54]

A strong EU, with the full cooperation of the member states, could
move in the direction of competitive transport and even a market in legal
pipeline entitlements. It would require the unambiguous definition and
allotment to shippers of pipeline capacity rights, the imposition of meticulous accounting standards, the constitutional definition of pipeline
property, the synchronization of pipeline capacity across national bor-
ders within the EU, a single regulatory jurisdiction for the major trunk
gas pipeline companies, a vastly enhanced market information system
to make transparent what is now opaque, the structural separation of
pipeline companies, and the imposition of the commodities clause. Even
then, it seems not yet certain whether in Europe, with its civil codes, the
creation of tradable property that comprises a bundle of legal entitlements is readily possible. Bundling the collection of legal rights into the
kind of property that supports competitive pipeline transport would be a
job for Europe's legal scholars.

Splitting legal transport entitlements from pipeline ownership can
be expected to cause dislocations, price spikes, and other new forms of
market abuse—as occurred in the United States. A competitive pipeline transport market may leave some pipelines stranded and others constrained and in need of expansion. Regulators and gas buyers would have
to brace themselves for the market's shakeout. The EU will also have
to deal with the long-term, oil-indexed gas contracts that now dominate
that system (as they once dominated the gas market in North America).

To say that all of this would pose a challenging administrative, legal,
and political task would be an understatement. Those who hold market
power (whether investor-owned or state-owned firms, or gas-supplying
nations) bitterly resist giving it up. Yet removing pipeline market power
is key to competitive gas transport. The payoff for the EU would be
great, as it has been in North America, where at every point in the continent, gas has taken on the apolitical character of other competitively
traded commodities. Indeed, as of early 2011, apart from the cost of
transport or distribution pipelines, the 115 million EU gas consumers
are paying roughly twice the price for their gas as their counterparts in
North America—representing a difference of well more than €50 billion
annually—for a variety of reasons including the lack of a competitive
market for gas. Such a relative price tag for the EU's gas consumers may
again confirm Mancur Olson's view that there is no constraint on the

social cost that a narrow collective group (such as producers and allied pipelines) "will find it expedient to impose on the society in the course of obtaining a larger share of the social output for itself."[55] The EU is in for some decades of work on the institutions that govern the way its pipelines are regulated, transact with customers, facilitate or impede the growth of a competitive gas market, and promote the security of its gas supplies without resorting to redundant pipelines and complex political maneuverings.

CHAPTER NINE

Making Sense of Pipelines

The Lenses of the New Institutional Economics

Adopting the proper economic perspective means everything in making sense of pipelines. Oliver Williamson identified in his 2009 Nobel Prize address why the neoclassical tradition was so ill matched to analyzing and understanding certain industrial relationships: it employs the *lens of choice* rather than the *lens of contract*.[1] The lens of contract illuminates asset specificity and the unsuitability of common carriage for organizing the pipeline industry. The lens of contract leads policy makers to strengthen the institutions needed to support long-term pipeline agreements of sufficient certainty to motivate pipeline investments. The lens of contract clarifies much that the neoclassical lens of choice overlooks about the governance of pipelines and the possibilities for competitive inland transport.

The pipeline industry is unlike other regulated businesses. Long-distance pipelines are neither public utility monopolies (like local energy or water distribution companies) nor transporters traditionally subject to price regulation but better left unregulated (like airlines or motor transport firms). Long-distance pipelines, burdened by asset specificity, present special problems of financing and contracting not encountered elsewhere. Applying traditional utility regulation to pipelines eliminates the prospect for competitive entry—stunting commodity markets. But deregulation of pipeline prices would leave customers geographically captive to a single pipeline subject to abuse. This volume shows that pipeline transport can be competitive in construction and use and the vehicle for unfettered fuel market competition. But it also shows that the institutional foundation for such competitive transport is at once inflexible (as in the definition of regulated pipeline property) and nuanced (as in the ability of mature and independent regulators to let go of some traditional levers of regulation to grasp others). Indeed, a century of dealing with oil and gas pipelines shows just how hard it is to keep them from being used as John D. Rockefeller first discovered they could be—as levers to frustrate competition in commodity markets and as profitable tollgates lying athwart commodity trade routes. Remedies for such familiar pipeline problems have been hard won. The early attempts in the United States to limit the power of oil pipelines to concentrate commodity markets, or to exact excessive tolls, were a lasting failure. Drawing upon experience in rail regulation, those attempts only diffused or transferred that power because of the legitimate need for pipeline investors to protect the value of their long-lived and immobile capital assets. Recent attempts in Europe or Australia to strengthen pipeline regulation have been no more successful. By using regulatory models that are also ill suited to independent pipeline transport, those attempts have done more to create a state-based patchwork of opaque monopolies across those continents than to promote competitive pipeline transport and independent gas markets.

Much of this volume examines the story of how Congress and federal regulators learned over the decades to create meticulous and reliable governance institutions suited to pipelines. It was a group effort, combining reliable, legislated accounting rules, competent regulators, strong interest groups of pipeline users, and wise judicial decisions. The type of competition in intangible property rights that Ronald Coase foresaw in 1960 was not easy to impose upon an existing, investor-owned pipeline system in a country with long-standing institutions geared to prevent the marauding of the value of private property by legislatures or regulatory agencies. The idea of creating similarly competitive regimes on other pipeline systems—particularly for gas users in Europe, where pipeline transport competition would appear to have such promise—seems harder yet. Distressingly, much of the analysis and policy purporting to reflect economic theory, or wise utility pricing, has not helped.

Modern economists have done little to make sense of the industry, to place it in a conceptual economic framework from which to craft useful public policy. Much of the insightful work that went into the economic institutions upon which competitive pipeline transport now rests in the

United States was done by institutional economists in the first four decades of the twentieth century, prior to the birth of the neoclassical tradition. More was accomplished by farsighted politicians and jurists. It was an astute US senator who at a critical moment saved the gas pipeline system from the stultifying effects of nineteenth-century notions of common carriage. Through the three decades that common carriage as a governing institution was held at bay, the states, Congress, and the courts defined the individual regulatory elements that could make contract-based, transcontinental pipeline transport work. Thus, when public opinion demanded an end to the regulatory vacuum in which interstate gas pipelines lived, Congress had the tools to craft pathbreaking legislation sensitive to the particular concerns of collective groups of fuel producers, financers, and pipeline users. And in the inevitable legal challenge, the Supreme Court provided an objective definition of the value of pipeline property—in a ruling called by economists of the era, without hyperbole, one of the most important pronouncements in the history of American law. All of this regulatory, legislative, and judicial effort happened in the first half of the twentieth century—just before the neoclassical tradition began and decades before the latter-day institutionalists refocused economic analysis on transaction costs, governance institutions, and public choice.

The final transition to competitive pipeline transport late in the second half of the century was accomplished by an insightful regulatory commission, pushed by powerful interest groups of gas distributors and their advocates. Economists of the neoclassical tradition who wrote about pipelines during this era generally either were preoccupied with economies of scale (which counts for little in the larger view of the industry), wrongly accepted *institutional barriers*—such as common carriage—as *natural barriers* that would apply to all inland transport, or abstracted from transport markets to focus on commodities. Overall, neither the institutional foundation nor the final transition to competitive pipeline transport in the United States owes anything to the neoclassical economic tradition—contrary to other areas of regulatory reform, such as in airline transport, for which that perspective is better suited.

In many respects, the application of neoclassical economic analysis to pipelines has been counterproductive. Most of what has passed for hopeful innovation in regulated pricing among the world's privatized pipelines (e.g., marginal-cost-based pipeline pricing models, pipeline capacity auctions, half-hourly spot markets for pipeline service) ignores asset specificity and frustrates the prospects for reliable contracting. The result of such efforts is large-scale pipeline monopolies immune from the threat of entry, perhaps coupled with "independent system operators" to deal with the inevitable desire among pipeline firms to remain vertically integrated. And for their part, those independent system operators can be predicted naturally to ally with the protectionist forces seeking to retain market power and freedom from the pressure of competitive entry.

Viewed through Williamson's lens of contract, a century of otherwise seemingly ad hoc elements of pipeline regulation and development comes into focus. Two elements of that perspective help. The first involves *property*: who provides the capital (investors or the public) and, if investor capital is involved, whether the value of that property is sufficiently well defined to permit predictable regulated prices to allow investors to be assured that the opportunity cost of their funds will be repaid over the useful lives of the assets involved. The second is *common carriage* or TPA: whether it is legal to organize the industry around contracts at all. Complementing the lens of contract are two other perspectives of the latter-day institutionalists that explain where existing institutions came from and whether they can reasonably be challenged: *collective action* (whether groups whose interest is advanced by the provision of competitive pipeline transport can press their case) and *institutional and political history* (whether the long-standing local customs or political boundaries encourage or disrupt the prospect for competitive transport).

The Value of Tangible Pipeline Property

Investor-owned pipelines existing in jurisdictions with vague definitions of the value of tangible regulated pipeline property have no practical ability to deal with asset specificity other than to foreclose competitive entry—either by evolving into large-scale, entry-protected utilities or through vertical integration. Among the world's pipelines, this question of whether pipeline property is cleanly and objectively defined is a sharp dividing line reflecting an evolutionary split more than a century old. The pipelines in the United States came first. At the start of the twentieth century, the utility executives who studied the public/private utility question tended to think that private financing was simply part of the American way of doing business. The reality was more complex, as the first major US inland transport projects—the early nineteenth-century canals—were publicly funded. Private funding only replaced public funding for such inland transport in America after the market perceived that private capital accumulation for railroads had a more certain prospect of allowing investors to recoup their capital. Undoubtedly, both long-standing tradition and the failure of public transport funding contributed to the acceptance of the idea that pipelines in the United States would be investor owned.

Investor ownership had critical consequences for industry organization and regulation. First, it sharply limited how Congress could act to regulate the industry, as any legislation perceived as damaging to private property without due process would ultimately be struck down by the US Supreme Court as unconstitutional (an issue that always shapes legislative debate in the United States). Second, it called for standardization of accounting and transparency so that the newly formed, independent, and specialized regulatory commissions could deal effectively and objectively with the underlying conflict between private enterprise and the public welfare. Third, the Supreme Court had to invent a definition for the value of regulated property (in its 1944 *Hope* decision) tailored to the unique valuation problems facing private regulated industries. The existence of private property—particularly in gas pipelines—drove these institutional advances, which subsequently allowed contracts to deal with asset specificity in an independent pipeline industry.

Without pressing needs to value tangible pipeline property for regulatory purposes, the institutions essential to support pipeline contracting do not arise. The US oil pipeline industry, hobbled by its maladapted— but still binding—1906 legislation, never succeeded under its indifferent regulatory commission in developing objective methods to value tangible pipeline investments. For pipelines built with public funds, the whole private property question is moot. For public pipeline projects, there was never any need to create accounting or regulatory systems to ensure that individual pipelines paid for themselves over their useful lives.

The consequence of this lack of a need to value tangible regulated pipeline property is that for the privatized pipelines of the world, an independent, investor-owned pipeline transport business does not have the basic tools to deal with asset specificity. Lacking the ability to contract reliably for the long lifetimes of the assets involved, pipeline undertakings need other governing arrangements. Some pipelines remain in public hands, where the question of tangible property values continues to be largely secondary and uncontroversial (or at least well buried in larger government budgets). Others, particularly those pipelines privatized as independent entities, are effectively regulated as franchised public utilities, which prevents competitive entry and the bypass of existing pipelines. Still others employ vertical integration to tie fuel supply, long-distance transport, and local distribution together into closed systems—also highly resistant to competitive entry.

The perspective of transaction cost economics, with its lens of contract, makes the organization of the world's pipeline sectors reasonably plain to see. Without the meticulous institutional foundation that contracting requires in the specialized case of tangible regulated property— freeing pipeline investors from fear of loss at the hands of either opportunistic regulators or opportunistic pipeline users—contract-based fuel transport on an independent, investor-owned pipeline system subject to competitive entry is impossible. Furthermore, no attention to regulated prices, service limits, auctions, or other detailed tariff-making or operational factors will help. There is no evolutionary path to competitive pipeline transport—through Coasian markets for intangible transport rights—that does not begin with settling the question of the value of tangible regulated pipeline property.

The Burden of Common Carriage and TPA

In the case of capital-intensive inland transport systems, the lens of contract highlights common carriage as a quaint but counterproductive nineteenth-century custom. While employing mild and seemingly inoffensive language about preventing discrimination, common carriage stops pipeline transacting in its tracks. The only sort of economic governance structures left for transport pipelines are vertical integration or the pipeline equivalent of a franchised public utility immune from the threat of entry.

The responsibilities implied by common carriage are so contrary to the needs of investors in pipelines that imposing them inevitably sets off a scramble to create other institutions to work around the restrictions. Oil pipelines in the United States created and applied a series of complicated and restrictive measures to limit access to preferred pipeline users soon after the ink was dry on its regulations. Ultimately, US oil companies found that only through vertical integration and widespread joint-venturing, in addition to those other restrictive measures, could the problems presented by "common" access to a business with such asset specificity be overcome. A century of such evolution has created such a complex oil pipeline industry—with such interlocked vertical and horizontal relationships and complex operating rules—that the question of applying the gas pipeline regulatory advances in the United States to oil pipelines never even comes up. Without the lens of contract, the governance schism between these two pipeline systems is a mystery arising out of the mists of a complex industrial history. The lens of contract pierces that mist to reveal the source of the resulting convoluted and generally uncompetitive—by gas pipeline standards—industry structure.

The imposition of TPA on the pipeline systems in Europe, and to a certain extent in Australia, compounds the difficulty in writing life-of-pipeline contracts that could handle asset specificity. For most of Europe's gas pipeline industry, TPA means that incumbent pipelines act like franchised public utilities. The major, large-scale new "interconnector" pipelines are exempt from TPA—it could hardly be otherwise given the huge investments involved. But the owners of those interconnectors are generally the incumbent pipelines or major gas suppliers, and hence vertical integration still helps to bind the parties together to support such projects (similar to vertically integrated joint ventures in US oil pipelines).

The lens of contract illuminates common carriage and TPA as inimical to the organization of an independent pipeline transport sector. The layers of complex pricing and service rules, in addition to the embrace of institutions that inherently disrupt competitive entry—such as "independent system operation"—merely entrench such institutions and make competitive inland fuel transport harder to achieve. Viewed through such a lens, no amount of fussing with regulated prices or other aspects of governance clears the competitive roadblock represented by the prohibition of long-term contract-based access.

The Lens of Collective Action

Layers of institutions—legislation, judicial precedent, regulations—surround pipeline systems. Unique to peculiar local circumstances, each layer is shaped by people who draw upon their own experience to craft solutions to particular social conflicts consistent with the apparent needs of the industry at particular points in time. But who describes those needs? The question for the development of pipeline markets is whether consumer pressure groups are able to stand eye to eye with small groups of pipelines or oil companies in counseling those who make the institutional and regulatory rules. Nothing distinguishes US and European gas pipeline systems more starkly than the existence of powerful pressure groups of independent distributors in the United States and their absence in Europe.

The creation of effective pressure groups of gas distributors in the United States, representing the interests of gas consumers, was attributable to Congress, which forcibly split gas transport from local distribution in what was rightly called by a contemporary economist the most stringent corrective legislation ever enacted against an American industry—a remedy suited to the patient. The remedy created pressure groups of well-funded local distributors who retained counsel and advisory economists through the second half of the twentieth century. Those pressure groups defeated attempts by oil and gas producers to deregulate wellhead prices prior to the creation of effective contract carriage on the US gas pipeline system. Later, those same pressure groups were instrumental in overcoming all of the various attempts by the interstate pipelines to retain elements of their traditional market power. Without the original radical surgery performed by Congress on an abusive and acquisitive industry of vertically integrated pipeline companies, it is unlikely that the persistent attempts by gas producers and pipelines to retain their sources of market power in the commodity pricing and inland transport of gas would have been so successfully overcome.

Vertical integration and the seemingly well-intentioned—but distinctly counterproductive—initiative of "full retail access" have denied to gas consumers the benefit of having well-funded advocacy groups to speak on their behalf. Without such advocates, those consumers are no match for vertically integrated national gas companies, or their respective national energy ministers. If Europe had independent gas distributors that acted on behalf of the continent's many millions of consumers, those consumers would face a fairer fight.

To be sure, even if independent gas distributors were to arise as a powerful pressure group on behalf of their millions of constituents in

Europe, the path forward to competitive gas transport would be a long and uncertain one. A plan to create a single EU pipeline regulator would require individual national regulators to cede jurisdiction over trans-European gas transportation—which is against the nature of any regulator. Creating a single publicly accessible accounting and information framework across the EU would require a level of transparency, for both business and operating records, that business managers or shareholders inherently do not want to provide. Proposals to bar incumbent pipelines from profitably trading in the gas they handle can be expected to motivate their strong opposition. Creating the conditions for the seamless cross-border use of the continent's gas pipelines is inconsistent with the incentives of those wedded to protecting their own vertically integrated "national champion" gas companies or to preventing the exposure of what a competitive transport market would label redundant pipelines.

Whether collective action on the part of consumer-oriented local gas distributors could overcome any of these institutional barriers—let alone all of them as Coasian bargaining would require—is a serious question. What Mancur Olson might have called the "lens of collective action"—or others today could call the "lens of public choice"—illuminates the prospects for encouraging rivalry in the organization of businesses wrapped in layers of such governing institutions. Without the purposeful creation of sophisticated advocates for fuel consumers—as happened both in the United States in the 1930s and in Argentina upon privatization—it is likely that fuel producers and their allied incumbent pipelines, for which competitive rivalry is unwelcome, will carry the day.

The Lens of Institutional History

The institutions that first arose to regulate pipelines were the products of social conflict between private enterprise and the public welfare. They came into being as people—legislators, judges, regulators—drew on the tools at hand to deal with immediate problems, thereby locking in institutional structures that subsequently would be viewed as "natural." It is hard to widen the perspectives of those who take as natural the institutions they grow up with. Such is a problem for the analysis of pipelines, with the thick layers of politics, public policy, and legislation surrounding them. Consider the question of *natural monopoly*. It is easy for those not disposed to view pipelines through the lenses of the new institutional economics to believe that a region's pipeline system is a natural monopoly when that assumption accompanied its monopoly planning and financing with public funds.

The institutional history of North America does not support such a natural monopoly view of pipelines. Since the nineteenth century, investor financing of transport projects, including railroads and pipelines, displayed elements of genuine competitive rivalry. Those who now study the pipeline industry's history there know that this mode of inland transport is a natural monopoly only in the narrowest sense. North American economists take for granted other institutions, formed many decades ago, dealing with legislated regulatory accounting, constitutionally protected property values, and administrative due process. The reverence with which those institutions are regarded by economists in both Canada and the United States tends to be a mystery to economists elsewhere.

For an industry of such reach, durability, and immobility, political boundaries are no less important than asset specificity in explaining institutional development and regulation. It is difficult to overstate the orderliness in the regulation of continental pipeline transport markets that flows from the US Constitution's *commerce clause,* which gives the federal government sole and unambiguous jurisdiction over pipelines used in interstate trade. The EU has no analogous institution that transcends the boundaries of its sovereign member states. Australia has only within recent memory moved some of the jurisdiction for pipeline regulation out of the hands of its states and into those of its federal authorities. Not having a single, authoritative regulator on a continental scale, for a continental pipeline transport system, muddies everything related to the prospect for either orderly regulation of the pipeline system or the promotion of competitive transport.

The lens of institutional history shows, as much as anything, that the evolution of institutions capable of handling such complex industries as major pipeline transport systems is at best quixotic and piecemeal. Not everything is up for grabs all the time. Indeed, very little is up for grabs most of the time. And even then, often what is up for debate is driven by public opinion, which works in unpredictable and complex ways. But a splendid Coasian script now exists for competitive pipeline transport, and perhaps the prospect for such a competitive advance elsewhere is greater because of it.

Pipelines and the Latter-Day Institutionalists

Pipelines are perhaps the ultimate industry by which to illustrate the power of the insights of the latter-day institutionalists. Pipelines sit in place for decades. There is little novel technology—and surely no romance—in pipelines as they have crossed the countryside transporting fuel from one point to another for over a century. The organization of the industry is necessarily about institutions of economic governance: public or private ownership, legal systems, political boundaries, and regulatory legislation. Economic theory that abstracts from such institutions has no chance of making sense of the organization of major pipeline systems or their competitive possibilities. Determining whence today's pipeline systems came, and where they are going, requires the multidisciplinary and historical approach of those who have fueled the resurgence of economic analysis based on governance institutions.

Pipeline transportation is an inherently difficult industry to study. The governing institutions are stubbornly dissimilar, highly complex, and often enough designed to frustrate rather than to facilitate economic analysis. Pipelines are literally buried far upstream from consumers, and the cost of the fuels they transport is figuratively buried in many millions of heating bills, in electricity prices, in manufactured goods, and in motor fuel. Those pipelines that hold market power—to set fuel prices or bar competitive entry into fuel transport—manifestly would prefer to exercise that power unseen. Only in the United States is pipeline accounting and operational information readily available, due to institutions imposed by the US Supreme Court in decades-old decisions. The secrecy that the industry works to maintain is demonstrated when European economists perform empirical pipeline studies not with European but with US pipeline data because those data are all they can get their hands on.

The study of international governing institutions is always difficult. But as the economic analysis of institutions has risen from almost total obscurity to prominence, since Coase, Williamson, Olson, North, and others made their contributions, it bodes better for the analysis of the pipeline industry and its competitive potential. Certainly for the study of a technologically straightforward industry whose evident diversity in worldwide organization and competitiveness owes to little else other than institutional diversity, it is the only kind of economic analysis that matters. Governance institutions explain why US oil and gas pipeline systems split, on the floor of the US Senate on May 4, 1906, onto different evolutionary paths. Such institutions explain why pipeline systems outside of North America grew up with little attention to the value of the property they represented. Institutions explain why it is easy to determine which pipeline capacity exists—precisely where and at what price—to ship across the continental United States but impossible to do the same across the continents of Europe or Australia. The extent to which competitive pipeline transport—and related competitive inland fuel markets—spreads is thus a question of whether local governing institutions are pushed to evolve to support such competition or are purposely maintained to prevent it.

Notes

Chapter One

1. Paul A. Samuelson and William D. Nordhaus, *Economics*, 12th ed. (New York: McGraw Hill, 1985), 526. Kahn reported to this writer that President Carter's original idea was merely to reform the regulation of the industry—loosening its apparent rigid cartelization. It was only after seven months as the chairman of the Civil Aeronautics Board that Kahn saw no plausible halfway house to complete deregulation. He used "marginal costs with wings" to signal his intention of reforming the industry in keeping with economic principles, uninfluenced by the romance of air travel.

2. See Clifford Winston, "Lessons from the U.S. Transport Deregulation Experience for Privatization" (discussion paper no. 2009–10, OECD/ITF Joint Transport Research Center, Paris, 2009); and Alfred E. Kahn, "Reflections of an Unwitting 'Political Entrepreneur,'" *Review of Network Economics* 7, no. 4 (Dec. 2008): 616–20. One of Kahn's fears—that the entire industry would fall under the control of a handful of multimarket operators dominating the flow of traffic originating and terminating at their respective hubs—never materialized.

3. That was the year when Paul Samuelson of the Massachusetts Institute of Technology (MIT) submitted an early version of his seminal book *Foundations of Economic Analysis* as his doctoral dissertation at Harvard. See Paul A. Samuelson, *Foundations of Economic Analysis* (Cambridge, MA: Harvard University Press, 1947).

4. Oliver Williamson summarizes his view of the approach taken by the new institutional economics: "The NIE has progressed not by advancing an overarching theory but by uncovering and explicating the microanalytic features to which [Kenneth] Arrow refers and by piling block upon block until the cumulative value added cannot be denied." Oliver E. Williamson, "The New Institutional Economics: Taking Stock, Looking Ahead," *Journal of Economic Literature* 38 (Sept. 2000): 596.

5. Kenneth Arrow posed the question: "Why did the older institutional-
ist school fail so miserably, though it contained such able analysts as Thorstein
Veblen, J. R. Commons, and W. C. Mitchell?" Arrow credited the latter-day institutionalists with merging economic history with sharper reasoning not simply to give new answers to the traditional questions of resource allocation but to
address new questions, such as why economic institutions emerged the way they
did and not otherwise. Kenneth J. Arrow, "Reflections on the Essays," in *Arrow
and the Foundations of the Theory of Economic Policy*, ed. George Feiwel (New
York: New York University Press, 1987), 734.

6. In his outgoing presidential address to the American Economic Association in 2009, Avinash Dixit, of Princeton University, concluded that the deepest
study of the origins and mechanisms of economic governance, which he defines
as "the structure and functioning of the legal and social institutions that support
economic activity," is the best source of advice on how to reform economic institutions and policies. Avinash Dixit, "Governance Institutions and Economic
Activity," *American Economic Review* 99, no. 1 (Mar. 2009): 5.

7. Ex post costs include "maladaptation and adjustment that arise when contract execution is misaligned as a result of gaps, errors, omissions, and unanticipated disturbances." Oliver E. Williamson, *The Mechanisms of Governance*
(New York: Oxford University Press, 1996), 379. Williamson was not the first
economist to grapple with transaction costs and their consequences. He cred-
its John R. Commons, of the University of Wisconsin, with that insight. See Oliver E. Williamson, *The Economic Institutions of Capitalism* (New York: Free
Press, 1985), 3.

8. "[Transaction cost economics] concurs that bounded rationality is the appropriate cognitive assumption and takes the chief lesson of bounded rationality for the study of contract to be that *all complex contracts are unavoidably incomplete*" (emphasis in the original). Oliver E. Williamson, "Transaction Cost
Economics," in *Handbook of New Institutional Economics*, ed. Claude Ménard
and Mary M. Shirley (Dordrecht, Netherlands: Springer, 2005), 46. The empirical studies have shown the relationship between contracting complexity and ex
post behavior. See, for example, Keith J. Crocker and Kenneth Reynolds, "The
Efficiency of Incomplete Contracts: An Empirical Analysis of Air Force Engine
Procurement," *RAND Journal of Economics* 24, no. 1 (Spring 1993): 126–46. It
is the incompleteness of contracts that, combined with the potential for ex post
opportunism, pushes organizations toward different forms of governance, and
which has been tested by numerous authors. Crocker and Masten survey the empirical literature, including the relationship between asset specificity and vertical
integration. Keith J. Crocker and Scott E. Masten, "Regulation and Administered Contracts Revisited: Lessons from Transaction-Cost Economics for Public
Utility Regulation," *Journal of Regulatory Economics* 8 (1996): 5–39.

9. According to George Stigler, "The world of zero transaction costs turns

out to be as strange as the physical world would be without friction." George J. Stigler, "The Law and Economics of Public Policy: A Plea to Scholars," *Journal of Legal Studies* 1 (1972): 12.

10. Eirik Furubotn and Rudolf Richter present a list of ten types of institutional arrangements that are taken as "allocatively neutral" in a neoclassical economic world without transaction costs. Eirik G. Furubotn and Rudolf Richter, *Institutions and Economic Theory: The Contributions of the New Institutional Economics* (Ann Arbor: University of Michigan Press, 1997), 9–10.

11. "Meanwhile, I was trying to find what could be the unit of investigation which would include these three constituents of conflict, dependence and order. After many years I worked out the conclusion that they were found combined together only in the formula of a *transaction*, as against the older concepts of commodities, labor, desires, individuals, and exchange." John R. Commons, *Institutional Economics* (New York: Macmillan, 1934), 4.

12. Ronald H. Coase, "The Nature of the Firm," *Economica* 4, no. 16 (1937): 386–405. As Paul Joskow points out, it was not an accepted part of received economic wisdom, until recently, that firms and markets were substitute governance mechanisms. Paul L. Joskow, "Vertical Integration," in *Handbook of New Institutional Economics*, ed. Claude Ménard and Mary M. Shirley (Dordrecht, Netherlands: Springer, 2005), 323.

13. Coase, "Nature of the Firm," 389. Coase was reacting to what he saw as a gap in existing economic thought, for at the time there was no acceptable theory to explain the existence of vertically integrated firms. In particular, he critiqued the well-known and highly regarded earlier work of Frank Knight, of the University of Chicago, who had wrestled with vertical or horizontal firm integration. Frank H. Knight, *Risk, Uncertainty and Profit* (1921; repr., Chicago: University of Chicago Press, 1971). Knight had concluded that it was impossible to "treat scientifically" the determinants of the size of a firm, holding that "the relation between efficiency and size [of firms] is one of the most serious problems of theory, being . . . largely a matter of personality and historical accident rather than of intelligible general principles." Coase, "Nature of the Firm," 394.

14. Williamson defines opportunism as "self-interest seeking with guile, to include calculated efforts to mislead, deceive, obfuscate, and otherwise confuse. Opportunism should be distinguished from simple self-interest seeking, according to which individuals play a game with fixed rules that they reliably obey." Williamson, *Mechanisms of Governance*, 378.

15. Indeed, the lesson from a century later seems to be that attempting to correct antitrust problems without taking into account the perspective of transaction cost economics for such an industry leads to unintended consequences. Williamson makes a similar point with respect to franchise bidding for cable TV: in an attempt to solve the natural monopoly problem through cable TV franchise bidding, policy makers overlooked ex post opportunism. See Oliver E. William-

son, "Franchise Bidding for Natural Monopolies—in General and with Respect to CATV," *Bell Journal of Economics* 7, no. 1 (Spring 1976): 73–104. Transaction cost economics lessons for antitrust are discussed in Paul L. Joskow, "Transaction Cost Economics, Antitrust Rules, and Remedies," *Journal of Law, Economics and Organization* 18, no. 1 (2002): 95–116.

16. See Susanne Lohman, "Rational Choice and Political Science," *The New Palgrave Dictionary of Economics*, ed. Steven N. Durlauf and Lawrence E. Blume, 2nd ed. (New York: Palgrave Macmillan, 2008), 5 of 6. Justice Oliver Wendell Holmes said in an address given in Boston in 1897: "Most of the things we do, we do for no better reason than that our fathers have done them or that our neighbors do them." Oliver Wendell Holmes, "The Path of the Law," address, 1897, reprinted in *Collected Legal Papers* (New York: Harcourt, Brace and Howe, 1920), 167, at 185.

17. Lance E. Davis and Douglass C. North, *Institutional Change and American Economic Growth* (Cambridge: Cambridge University Press, 1971), vii. In this book, which looked closely at the development of nineteenth-century transport systems in the United States, among other subjects, Davis and North said they made no pretense to present either a fully formed theory or a complete history of the transport systems involved (e.g., canals). As they said, their case studies "were written to illustrate the promise, and achievements, and the limitations of the theory" (ibid.).

18. North found that early nineteenth-century large-ship, low per-unit cost technology existed by 1600 in the Dutch "flute" ship. But its adoption outside of the relatively safe Baltic routes had to await the decline in the institutions of piracy and privateering on the oceans generally, which until the end of the Napoleonic Wars, had required faster, armed ships and thus higher per-unit shipping costs. Douglass C. North, "Sources of Productivity Change in Ocean Shipping," *Journal of Political Economy* 76 (1968): 953–70.

19. As Davis and North recount, this was a game-changing exercise of political opportunism on the part of legislators in New York. Thereafter, financiers preferred to lend to private interests for large-scale inland transport projects rather than to demonstrably opportunistic governments. Davis and North, *Institutional Change*, 140–43.

20. Ibid., 48–51, 135–66.

21. John R. Commons, *The Economics of Collective Action* (New York: Macmillan, 1950), 21.

22. Contract holders may hire agents to bundle the two services, without restrictions, if they wish to do so.

23. Ronald H. Coase, "The Federal Communications Commission," *Journal of Law and Economics* 2 (1959): 14.

24. Ronald H. Coase, "The Problem of Social Cost," *Journal of Law and Economics* 3 (1960): 1–44. In essence, the idea attributed to Coase states that given

well-defined property rights, low transactions costs, perfect competition, complete information, and the absence of other barriers to efficient resource allocation, resources will be used efficiently regardless of who initially owned them, resolving all private externalities in the process. John Dales published an influential essay in 1968 concerning environmental markets that, as Coase had done, connected prices (what Dales called "the stuff of economics") and the law of property rights. See J. H. Dales, *Pollution, Property and Prices* (Toronto: University of Toronto Press, 1968).

25. Steven Cheung relates the story of how, in 1960, Coase convinced a highly skeptical group of eminent economists at the University of Chicago, including Milton Friedman, George Stigler, Arnold Harberger, Reuben Kessel, John McGee, Aaron Director, and others, that his theory about tradable property rights worked. Coase's theory seemed to contradict a long-held principle of early twentieth-century British economist Arthur Pigou. Steven N. S. Cheung, "Ronald Henry Coase (b. 1910)," *The New Palgrave: A Dictionary of Economics*, 1st ed. (London: Macmillan, 1987), 1:456. The criticisms Coase leveled at Pigou were perhaps unfair. Vic Goldberg states that Pigou had raised many of the issues that Coase and Williamson would later emphasize and that Coase's 1960 paper may well have conformed to Pigou's own understanding of an approach to problems of market failure. See Victor P. Goldberg, "Pigou on Complex Contracts and Welfare Economics," *Research in Law and Economics* 39 (1981): 42–43.

26. See A. Denny Ellerman, Paul L. Joskow, and David Harrison Jr., "Emissions Trading in the US: Experience, Lessons and Considerations for Greenhouse Gases" (Pew Center on Global Climate Change, May 2003); and Evan R. Kwerel and Gregory L. Rosston, "An Insider's View of FCC Spectrum Auctions," *Journal of Regulatory Economics* 7, no. 3 (May 2000): 253–89.

27. But, to be sure, all US interstate gas pipelines continue to charge cost-based regulated prices.

28. Such is not true for point-to-point high-voltage DC power lines where separation from the interconnected grid permits the creation and sale of point-to-point power capacity transport rights.

29. Winston, "Lessons," 4–8.

30. Adam Smith, *The Wealth of Nations*, book 1, chapter 10, part 2 (New York: Modern Library, 1937), 142.

31. Mancur Olson, "Collective Action," *The New Palgrave Dictionary of Economics*, ed. Steven N. Durlauf and Lawrence E. Blume, 2nd ed. (New York: Palgrave Macmillan, 2008), 3 of 5. To be sure, Smith's insights were not ignored by everyone. James M. Landis, in his famous report to President John F. Kennedy in 1960, denounced the effectiveness of US regulation in the 1940s and 1950s (citing that producing interests were typically better organized and effective in bending regulations their way than consuming interests). James M. Landis, *Re-*

port on Regulatory Agencies to the President Elect, US Senate Committee on the Judiciary, 86th Cong., 2nd sess. (Washington, DC: Committee Print, 1960).

32. See Mancur Olson, *The Logic of Collective Action: Public Goods and the Theory of Groups* (Cambridge, MA: Harvard University Press, 1965). The book is still in print almost a half century later.

33. More recent contributors to the theoretical and empirical tie between institutions and economic growth recognize the inherent interrelationships among collective action, political power, the endogeneity of institutions, and the distribution of capital resources. See Daron Acemoglu, Simon Johnson, and James A. Robinson, "Institutions as a Fundamental Cause of Long-Run Growth," in *Handbook of Economic Growth*, ed. Philippe Aghion and Steven N. Durlauf, vol. 1A (Amsterdam: Elsevier, 2005), 386–472. Spiller and Liao provide a good summary of the literature from the new institutional economics (NIE) perspective: "Recently, however, there has been an increase in literature which examines the participation of interest groups in public policy making from a new NIE perspective. The distinguishing feature of the NIE approach, as it is understood today, is its emphasis on opening up the black box of decision making with reference to, among other things, understanding the rules and the play of the game." Pablo T. Spiller and Sanny Liao, "Buy, Lobby or Sue: Interest Groups' Participation in Policy Making: A Selective Survey," in *New Institutional Economics— A Guidebook*, ed. Eric Brousseau and Jean-Michel Glachant (Cambridge: Cambridge University Press, 2008), 307.

Chapter Two

1. Stuart Daggett wrote an impressive book on the subject of rail and road transport early in the twentieth century. Theodore Keeler produced a similarly impressive study of railroads two generations later. See Stuart R. Daggett, *Principles of Inland Transportation* (New York: Harper and Brothers, 1928); and Theodore E. Keeler, *Railroads, Freight and Public Policy* (Washington, DC: Brookings Institution, 1983).

2. See Sally Hunt and Graham Shuttleworth, *Competition and Choice in Electricity* (New York: Wiley, 1996); and Jonathan E. Nuechterlein and Philip J. Weiser, *Digital Crossroads: American Telecommunications Policy in the Internet Age* (Cambridge, MA: MIT Press, 2007).

3. Indeed, that institutional complexity is a barrier to almost any observer of the world's oil and gas pipeline businesses. When delegations from Russia's Transneft—or South Africa's Petronet—oil pipeline systems have traveled to the United States to learn about oil pipeline regulation, they returned confused over the century-long history of the industry's structure, pricing, and complicated shipping arrangements (as I heard myself, having dealt with staff from

both firms). And the decades-long evolution of institutions surrounding the US gas pipeline system seems so irrelevant to the current problems of European gas markets, and the perceived threat to their security from Russia, that its development and current status seem to many in Europe to merit little careful analysis.

4. Paul W. MacAvoy, *Price Formation in Natural Gas Fields* (New Haven, CT: Yale University Press, 1962); Paul W. MacAvoy and Stephen G. Breyer, *Energy Regulation by the Federal Power Commission* (Washington, DC: Brookings Institution, 1974); and Paul W. MacAvoy, *The Natural Gas Market: Sixty Years of Regulation and Deregulation* (New Haven, CT: Yale University Press, 2000). Morris Adelman wrote his own book on the oil industry, but like MacAvoy's, it only tangentially dealt with pipelines. Morris A. Adelman, *The World Petroleum Market* (Baltimore: Johns Hopkins University Press, 1972).

5. Arlon R. Tussing and Connie C. Barlow, *The Natural Gas Industry: Evolution, Structure and Economics* (Cambridge, MA: Ballinger, 1984); and Arlon R. Tussing and Bob Tippee, *The Natural Gas Industry: Evolution, Structure and Economics*, 2nd ed. (Tulsa, OK: PennWell Books, 1995).

6. Malcolm W. H. Peebles, *Evolution of the Gas Industry* (London: Macmillan, 1980).

7. Robert Mabro and Ian Wybrew-Bond, eds., *Gas to Europe: The Strategies of Four Major Suppliers* (Oxford, UK: Oxford University Press, 1999). The volume contains papers by Peebles and Jonathan Stern, of Oxford University (who analyzes the origin of Russia's gas export strategy).

8. Arthur M. Johnson, *The Development of American Petroleum Pipelines: A Study in Private Enterprise and Public Policy, 1862–1906* (Ithaca, NY: Cornell University Press, 1956); and Arthur M. Johnson, *Petroleum Pipelines and Public Policy, 1906–1959* (Cambridge, MA: Harvard University Press, 1967). As Johnson states in the preface to his second book, a number of the major oil pipeline companies in the United States admired his first book and agreed to underwrite a grant to Harvard Business School to produce the second, opening their company records for Johnson's examination. Johnson stressed that several pipeline representatives made factual or editorial suggestions, but in no instance did they interfere with his approach or conclusions.

9. Christopher J. Castaneda, *Regulated Enterprise: Natural Gas Pipelines and Northeastern Markets, 1938–1954* (Columbus: Ohio State University Press, 1993); Christopher J. Castaneda and Clarance M. Smith, *Gas Pipelines and the Emergence of America's Regulatory State: A History of Panhandle Eastern Corporation, 1928–1993* (Cambridge, UK: Cambridge University Press, 1996); and Christopher J. Castaneda, *Invisible Fuel: Manufactured and Natural Gas in America, 1800–2000* (New York: Twayne Publishers, 1999). Castaneda had obtained private corporate records from the Texas Eastern Pipeline for his first book and from the Panhandle Eastern Pipe Line for his second.

10. Eugene V. Rostow, *A National Policy for the Oil Industry* (New Haven,

CT: Yale University Press, 1948); and Eugene V. Rostow and Arthur S. Sachs, "Entry into the Oil Refining Business: Vertical Integration Re-examined," *Yale Law Journal* 61 (1952): 856–914.

11. George S. Wolbert Jr., *American Pipe Lines: Their Industrial Structure, Economic Status and Legal Implications* (Norman: University of Oklahoma Press, 1951); and George S. Wolbert Jr., *U.S. Oil Pipe Lines* (Washington, DC: American Petroleum Institute, 1979). Wolbert wrote the second book after joining Shell Oil. He was evidently involved in litigating every major issue regarding oil pipelines during his tenure at Shell, and his books read like extended legal briefs. He doggedly defends vertically integrated oil pipeline undertakings. But apart from the evident partisanship of his later book in particular, his work is useful in its encyclopedic recounting (with thousands of footnote citations to the positions of parties) of the policy struggles oil pipelines faced during the 1950s through the 1970s.

12. Richard J. Pierce, "Reconstituting the Natural Gas Industry, from the Wellhead to the Burnertip," *Energy Law Journal* 9, no. 1 (1988): 1–57.

13. M. Elizabeth Sanders, *The Regulation of Natural Gas: Policy and Politics, 1938–1978* (Philadelphia: Temple University Press, 1981).

14. Emery Troxel, *Economics of Public Utilities* (New York: Rinehart and Company, 1947); Emery Troxel, "Long-Distance Natural Gas Pipe Lines," *Journal of Land and Public Utility Economics* (November 1936): 344–54; Emery Troxel, "II. Regulation of Interstate Movements of Natural Gas," *Journal of Land and Public Utility Economics* (February 1937): 20–30; and Emery Troxel, "III. Some Problems in State Regulation of Natural Gas Utilities," *Journal of Land and Public Utility Economics* (February 1937): 188–203.

15. Leslie Cookenboo Jr., *Crude Oil Pipe Lines and Competition in the Oil Industry* (Cambridge, MA: Harvard University Press, 1955).

16. Cookenboo's first recommendation for public policy related to pipelines is that "[any] public policy for crude oil trunk pipe lines should ensure . . . [t]hat conglomerate mass transportation will be used to the greatest possible extent in order to achieve minimum costs for all companies." Cookenboo, *Crude Oil Pipe Lines*, 167.

17. Alfred E. Kahn, *The Economics of Regulation: Principles and Institutions*, vol. 2, *Institutional Issues* (New York: Wiley, 1971).

18. Davis and North, *Institutional Change*, 77–79, 139–43. Even the state of New York had trouble raising the $7 million for a canal that was 363 miles long, 20 feet wide, and 4 feet deep, with a rise of 630 feet and a drop of 62 feet from the Hudson River to Lake Erie.

19. *Municipal and Private Operation of Public Utilities*, 3 vols. (New York: National Civic Federation, 1907). The National Civic Federation investigating committee spent six months intensively studying dozens of publicly owned and

investor-owned utilities in the United States and United Kingdom with the goal of settling the issue of whether private or public ownership was in the nation's best interest. In recommending against public ownership, the committee's report helped shape a future of investor-owned utilities in the United States. The importance of this report in laying the foundation for private utility ownership, in a methodical and disinterested manner, was well recognized at the time. See William B. Munro, "Review: The Civic Federation Report on Public Ownership," *Quarterly Journal of Economics* 23, no. 1 (1908): 161–74. For an insightful personal recollection of the work of the Committee on Investigation that produced the report, for which Commons wrote that "the newspapers gave almost daily reports of the progress of our joint investigating committee," see John R. Commons, *Myself* (New York: Macmillan, 1934), 111–20.

20. Walton Clark, of the gas utility in Philadelphia and a member of the Committee on Investigation, appeared to sum up the opinion of those favoring private ownership with government regulation, saying: "Any government that is too feeble or corrupt to control with justice the conduct of a [privately owned] public service company, has little prospect of being able itself to supply such public service with efficiency and justice." *Municipal and Private Operation of Public Utilities*, 1:443.

21. Federal Power Commission v. Hope Natural Gas, 320 US 591 (1944). Canada has its own version of the *Hope* decision: Northwest Utilities v. City of Edmonton, S.C.R. 186 (NUL 1929). The *Hope* decision is key to understanding US regulatory institutions. It receives a treatment consistent with that importance in chapter 7.

22. Indeed, Davis and North conclude that the ICC did more to consolidate the market power of railroad cartels in the late nineteenth century than to effectively restrain their market power, largely by applying common-law rules of carriage that prohibited railroads from pursuing justifiable price discrimination or cutting prices to preserve traffic on underutilized routes. Davis and North, *Institutional Change*, 135–66.

23. A brief essay of mine appeared in 1993 describing the potential for Coasian bargaining in US regulated pipeline capacity, seven years before the final details of the new market were worked out by the federal regulator. "Gas Pipeline Capacity: Who Owns It, Who Profits, Who Pays?," *Public Utilities Fortnightly* 132, no. 18 (Oct. 1, 1993): 17–20. To an industry and regulator burdened by practical problems and pressure groups, my discussion must have seemed highly esoteric.

24. Coase overturned the notion, promulgated by the earlier British economist Arthur Pigou (or, perhaps more fairly, by his followers), that governments had the responsibility to hold back the potential developers who would provide society less value than others for the use of finite resources.

Chapter 3

1. For example, Leslie Cookenboo, of Rice University, recommended in his 1955 study of American oil pipelines that "conglomerate mass transportation" consisting of compulsory joint ventures be used to the greatest extent possible. *Crude Oil Pipe Lines*, 167–68.

2. See MacAvoy, *Natural Gas Market*, 99. Earlier, in joint work with Justice Stephen Breyer, MacAvoy maintained that the value of federal gas pipeline regulation "has been either very low or zero," largely because of what he held to be the absence of natural monopoly. *Energy Regulation*, 54.

3. There are some similar arguments in Thomas J. DiLorenzo, "The Myth of Natural Monopoly," *Review of Austrian Economics* 9, no. 2 (1996): 43–58.

4. In the past, there were practical limits on manufactured diameters for longitudinal seam-welded gas and oil pipelines—generally sixty-four inches. More recent advances in spiral welding, in which the pipe is made from a flat rolled steel sheet, formed into a pipe by spiraling, has produced diameters larger than sixty-four inches. See John L. Kennedy, *Oil and Gas Pipeline Fundamentals* (Tulsa, OK: PennWell Books, 1993), 50–60.

5. See William W. Sharkey, *The Theory of Natural Monopoly* (Cambridge: Cambridge University Press, 1982); and William J. Baumol, John C. Panzar, and Robert D. Willig, *Contestable Markets and the Theory of Industrial Structure* (New York: Harcourt Brace Jovanovich, 1982).

6. Baumol, Panzar, and Willig, *Contestable Markets*, 17.

7. For a single-product firm—for example, a pipeline company—the concept of a subadditive cost function includes little more than the straightforward attributes of declining average or marginal cost. For multiproduct firms, however, the concepts of natural monopoly and subadditive costs are more complex. While the investigation of real-world examples was not the point of the work of Baumol and his colleagues, the driver for that research, which was financed in large part by AT&T prior to its breakup in 1984, was the widest generalization of the definitions of natural monopoly that could be used to investigate the sustainability of the incumbent multiproduct firm's structure when competitors were permitted to enter and supply individual products within the multiproduct telecommunications industry.

8. James M. Henderson and Richard E. Quandt, *Microeconomic Theory: A Mathematical Approach*, 2nd ed. (New York: McGraw-Hill, 1971), 80–86. Marc Nerlove in 1963 was the first to apply duality theory to investigate the underlying production technology of an industry through the specification of a cost function—a technique that became popular for subsequent empirical work on the structure of costs in many industries. Nevertheless, he was bound by functional forms that assumed declining costs with scale. Marc Nerlove, "Returns to

Scale in Electricity Supply," in *Measurement in Economics*, by Carl F. Christ et al. (Stanford, CA.: Stanford University Press, 1963).

9. Laurits R. Christensen and William H. Greene, "Economies of Scale in U.S. Electric Power Generation," *Journal of Political Economy* 84, no. 4 (Aug. 1976): 655–76.

10. Johnson, *Petroleum Pipelines and Public Policy*, 146–48.

11. Troxel, "Long-Distance Natural Gas Pipe Lines."

12. The Uniform System of Accounts was applied by the Federal Power Commission to gas pipelines in 1938 (upon passage of the Natural Gas Act) and to oil pipelines in 1978 (upon passage of the Energy Policy Act of 1978, which transferred jurisdiction over oil pipelines from the Interstate Commerce Commission to the Federal Energy Regulatory Commission).

13. Troxel, "Long-Distance Natural Gas Pipe Lines," 349.

14. Alfred M. Leeston, John A. Crichton, and John C. Jacobs, *The Dynamic Natural Gas Industry* (Norman: University of Oklahoma Press, 1963), 83–85.

15. E. W. McAllister, ed., *Pipeline Rules of Thumb Handbook*, 4th ed. (Houston: Gulf Publishing Company, 1998), 510, 544.

16. Baumol, Panzar, and Willig, *Contestable Markets*, 408–13.

17. Ibid., 425–29. As they put it, "A fundamental limitation of sustainability analysis is its implication that entrants' expectations are those of the Bertrand-Nash models," without which the threat of entry may "evaporate even before it can be noticed" (428). That is, these models draw from the restrictive elements of game theory in·which the actions of incumbent firms depend on the nature and credibility of the threat of entry, leading to a wide set of possible outcomes depending on assumptions regarding how the incumbent firm will react, on the quite wide Bertrand-Nash continuum, to that threat. See Joseph Bertrand, "Review of *Théorie mathematique de la richesse sociale* and *Recherches sur les principles mathematique de la theorie des richesses*," *Journal des Savants* (1883): 499–508; and John F. Nash Jr., "The Bargaining Problem," *Econometrica* 18 (1950): 155–62.

18. For example, see Elizabeth E. Bailey, David R. Graham, and Daniel P. Kaplan, *Deregulating the Airlines* (Cambridge, MA: MIT Press, 1985), 166–71.

19. Christopher Castaneda provides a number of examples in the history of natural gas development in the United States. *Invisible Fuel*.

20. There were no such major gas pipelines in Europe before the 3,763-kilometer Russian Transgas pipeline reached Western Europe in 1972. Australia built no major gas pipelines until the 1,900-kilometer Moomba-to-Sydney pipeline in 1976. Mexico's Pemex (the country's constitutional oil and gas monopoly) began the substantial construction of its gas pipeline system, still smaller than Argentina's, in the 1950s. The TransCanada pipeline went into operation in 1957.

21. As part of the analysis leading to the privatization of Gas del Estado in 1991, I performed those "netback" calculations for the Argentine privatization authorities that led to a decision to keep the Austral basin pipeline operating by blending its asset base with that of the newer, shorter Neuquén pipelines.

22. According to a 1995 Department of Agriculture, Energy and Minerals assessment, proven reserves in the Bass Strait (Gippsland basin) were 8,280 petajoules, versus 2,800 for the Cooper basin.

23. The intended purpose of certificates of public convenience and necessity is generally to allow regulators to regulate competition in high-fixed-cost businesses. The process of obtaining a certificate is intended to show a regulator that there is a need for the investment and that the provider is fit, willing, and able to render the intended services.

24. See Kahn, *Economics of Regulation*, 2:152–71.

25. More often than not, the full commission largely adopted the view of the administrative law judges in the cases cited by Kahn.

26. For example, Kahn refers to the discussion of Judge J. Skelly Wright, quoting him at length: "Admittedly, the Commission possesses a rate-making power and this power is designed to protect the consumers of natural gas. But it is clear that this power is largely a negative one. . . . On the other hand, if competition exists, albeit in a limited area, there would be incentives for innovation by the regulated companies themselves and for their coming forward with proposals for better services, lower prices, or both." Kahn, *Economics of Regulation*, 2:161–62.

27. Kahn relates the comments of a reviewer of his book that reflect on the administrative burden confronting such entrants in the face of determined incumbents: "I shudder to think of what kind of [legal] record an imaginative and obstructive lawyer or group of lawyers might compile on that subject. . . . I would hate to undertake the task of proving any pipeline to be inefficient and unreasonably 'high cost' in its operating and then also prove that its present plight was not of its own doing before I could get a less expensive supply of gas from some other pipeline." Ibid., 168.

28. There are two noteworthy cases in which competitive rivalry in major pipeline expansions, to the East and West Coasts of the United States, overwhelmed the desire of regulators to pursue single projects based on notions of economies of scale. The first came in the late 1940s during the postwar expansion of gas pipelines into the New England region. The second involved capacity expansions into Southern California in the 1960s. The struggle between the concept of natural monopoly and the encouragement of competitive rivalry dominated both the headlines and the Federal Power Commission (FPC) proceedings during these expansions. Pipeline companies came before the FPC with competing plans to extend pipelines from existing fields to new markets. The FPC's administrative law judges, and then the FPC itself, faced a difficult choice

between apparent efficiency and competitive rivalry, as well as between long-term plans and practical expediency in the face of rapidly growing demand. In both cases, competitive rivalry won out, and at least two pipeline suppliers were chosen in both cases. For the New England case, see Castaneda, *Regulated Enterprise*, 145–50. For the Southern California expansions, see Kahn, *Economics of Regulation*, 2:154–57.

29. Of course, a pipeline has considerable market power in the short run in its immediate vicinity. Through price discrimination, an incumbent could exercise its market power in those areas. Despite the number of independent pipeline competitors in the United States and the robust market in the purchase and sale of capacity rights, that concern is the reason why the FERC has never relaxed cost-based price regulation on any of the US interstate gas pipelines under its jurisdiction.

30. James Nelson, of Amherst College, put the whole "natural" concept brilliantly into context:

> One of the most unfortunate phrases ever introduced into law or economics was the phrase "natural monopoly." Every monopoly is a product of public policy. No present monopoly, public or private, can be traced back through history in a pure form. . . . Roads? The "King's Highway" was usually more an easement than a facility until well into the eighteenth century, except where the admittedly monopoly-minded Romans had done their work; the highway was lifted from its literal morass only by private turnpike companies, sometimes on a quasi-competitive basis. . . . So "natural monopolies" in fact originated in response to a belief that some goal, or goals, of public policy would be advanced by encouraging or permitting a monopoly to be formed, and discouraging or forbidding future competition with this monopoly. (James R. Nelson, "The Role of Competition in Regulated Industries," *Antitrust Bulletin* 11 [1966]: 3)

Chapter 4

1. Wisconsin's statute was drafted by John R. Commons at the request of Governor Robert La Follette. New York's was largely the work of Governor Charles Evans Hughes, later chief justice of the US Supreme Court. See Commons, *Myself*, 120–28; and Merlo J. Pusey, *Charles Evans Hughes* (New York: Macmillan, 1952), 201–9.

2. National Civic Federation, *Municipal and Private Operation of Public Utilities*, 3 vols. (New York: National Civic Federation, 1907). Two economists who were important in the drafting of Wisconsin's and New York's regulatory statutes

(John R. Commons from Wisconsin and Milo R. Maltbie from New York) were prominent members of that National Civic Federation study, along with lawyer Louis Brandeis from Boston, the future Supreme Court justice. For the story of Maltbie's role in establishing regulation in New York (including Commons's role in Maltbie's appointment to the new Public Service Commission), see Howard J. Read, *Defending the Public: Milo R. Maltbie and Utility Regulation in New York* (Pittsburgh: Dorrance Publishing, 1998).

3. Commons taught the class based on the experience he gained as part of the National Civic Federation study. See Leonard O. Weiss, "The Field of Industrial Organization at Wisconsin," in *Economists at Wisconsin*, ed. Robert J. Lampman (Madison: Board of Regents of the University of Wisconsin System, 1993), 219.

4. See Keeler, *Railroads, Freight and Public Policy*, 22–24.

5. See Johnson, *Petroleum Pipelines and Public Policy*, 23–24.

6. By 1906, the states with sizable oil pipeline activity had conferred the right of eminent domain on pipelines. Common carriage on those lines was the expected quid pro quo requirement for aiding pipelines for what otherwise would mean dealing bilaterally with all property owners along their routes. See Johnson, *Petroleum Pipelines and Public Policy*, 20–21.

7. Cong. Rec., 59th Cong., 1st Sess., S6365 (May 4, 1906).

8. Johnson, *Petroleum Pipelines and Public Policy*, 21–22.

9. A description of that vertically integrated way of life in the oil industry early in the twentieth century appears in Melvin G. de Chazeau and Alfred E. Kahn, *Integration and Competition in the Petroleum Industry* (New Haven, CT: Yale University Press, 1959), 83–86 ("Integration as a Way of Business Life before 1911").

10. Garfield was the son of the assassinated president James Garfield and a member of President Roosevelt's "tennis cabinet," an informal group of people whom Roosevelt trusted in matters of state and whose company he enjoyed.

11. *Report of the Commissioner of Corporations*, xx.

12. Investigative journalist Ida Tarbell virtually reinvented investigative reporting with her history of the Standard Oil Company, which was serialized in nineteen installments by *McClure's* magazine between 1902 and 1904 before being published in book form. See Ida Tarbell, *The History of the Standard Oil Company* (New York: McClure, Phillips, and Co., 1904).

13. See 34 U.S. Stat. 584 (1906).

14. Tussing and Barlow, *Natural Gas Industry*, 29, 34.

15. Per the commerce clause of the US Constitution (article 1, section 8, clause 3), interstate commerce fell solely under the jurisdiction of the federal government.

16. West v. Kansas Natural Gas Co., 221 U.S. 229 (1910).

17. "The transportation, sale and delivery constitute an unbroken chain,

fundamentally interstate from beginning to end, and of such continuity as to amount to an established course of business. The paramount interest is not local but national—admitting of and requiring uniformity of regulation. Such uniformity, *even though it be the uniformity of governmental non-action*, may be highly necessary to preserve equality of opportunity and treatment among the various communities and states concerned." Barrett v. Kansas National Gas Co., 265 U.S. 298 (P.U.R. 1924 E78) (emphasis added). Troxel presents a good discussion of all these cases in the second of three survey articles on the gas pipeline industry he wrote in 1936 and 1937: "II. Regulation of Interstate Movements of Natural Gas," 21–22.

18. With the states powerless to regulate interstate gas pipelines, the question still remained where the responsibility of the interstate transporter ended and that of the local distributor began. The Supreme Court provided such a distinction in 1931: East Ohio Gas Co. v. Tax Com. of Ohio, 283 U.S. 465 (1931). This decision established the point within which states could exercise their jurisdiction over retail gas prices.

19. Western Distributing Co. v. Pub. Serv. Com. of Kansas, 52 S. Ct. (283 P.U.R. 1932 B 236).

20. See Troxel, "II. Regulation of Interstate Movements of Natural Gas," 25–26.

21. Castaneda, *Invisible Fuel*, 106. The 1928 study was not the only congressional investigation that made recommendations with respect either to holding companies or to the gas pipeline industry. In his report on oil pipelines for the House Committee on Interstate and Foreign Commerce, University of Texas economist Walter Splawn's first recommendation was that any regulation aimed at any corporation would have to "reach the owning or controlling corporations." Splawn also recommended at that time (in 1933) that "transportation of gas in interstate commerce by pipe line be regulated." See Walter M. W. Splawn, *Report on Pipe Lines* (in two parts), H.R. Rep. No. 2192, 72nd Congress, 2nd sess. (Washington, DC: US Government Printing Office, 1933), pp. 1:lxxvii–lxxix (the "Splawn report").

22. Natural Gas Act of 1938, 52 Stat., pp. 821–33. The act was approved on June 21, 1938.

23. "Energy and Transport in Figures," as presented in Paul Belkin, "CRS Report for Congress: The European Union's Energy Security Challenges," *Statistical Pocket Book, 2007* (Washington, DC: Congressional Research Service, January 30, 2008), 6.

24. Franziska Holz, Christian von Hirschhausen, and Claudia Kemfert, "A Strategic Model of European Gas Supply" (discussion paper 551, DIW Berlin, Jan. 2006).

25. As of the end of 2010, the annual volume of forward gas trades in the United States (17,743,018 million cubic meters) on the NYMEX market was

865 times the level of such trades in the markets of continental Organization for Economic Cooperation and Development (OECD) Europe: the Dutch TTF (17,741 million cubic meters) and the NCG and GASPOOL (2,773 million cubic meters combined). Considering the 2009 consumption volumes (646,347 million cubic meters for the United States and 436,441 million cubic meters for OECD Europe excluding the United Kingdom), this is a stark differential. More such trading takes place in the United Kingdom, which facilitated the trade by abstracting from transport through its national balancing point (348,350 million cubic meters volume against 90,495 million cubic meters consumption). Consumption data were obtained from the International Energy Agency. Futures volume data were obtained from CME Group (NYMEX), APX-ENDEX (Dutch TTF), the European Energy Exchange (NCG and GASPOOL), and the Intercontinental Exchange (UK NBP).

26. Some European gas pipeline companies (particularly in Germany and the Netherlands) are not state owned but include large municipal shareholdings.

27. The EU does have a Single Market Treaty, and EU law is supposed to override national law. However, the EU has no independent political mandate and so is dependent upon the agreement of member states for all new laws and regulations. The EU can exhort member states to do their best to regulate effectively and cooperate, but it lacks the power to dictate regulatory standards in member states or to overturn decisions of national regulators.

28. See Mark Armstrong, Simon Cowan, and John Vickers, *Regulatory Reform, Economic Analysis and the British Experience* (Cambridge, MA: MIT Press, 1994), 245–78; John Vickers and George K. Yarrow, *Privatization: An Economic Analysis* (Cambridge, MA: MIT Press, 1988), 245–54; and Peebles, *Evolution of the Gas Industry.* Also see David M. Newbery, *Privatization, Restructuring and Regulation of Network Utilities,* The Walras-Pareto Lectures, 1995 (Cambridge, MA: MIT Press, 2000).

29. There are, of course, important institutional constraints that prevent the easy breakup of integrated companies such as British Gas. In the 1990s, another British Gas board member (Christopher Brierley, whose acquaintance I made in working with gas pipelines under World Bank projects in Poland and Chile) told me that at the time of privatization, the chairman of the publicly owned company, Sir Denis Rooke, refused to cooperate with a government privatization effort that would have dismembered "his" gas company. He struck a deal with Peter Walker, energy minister at the time, in which he would cooperate with privatization as long as the gas transport, distribution, and retailing business was turned, by act of Parliament, from a publicly owned monopoly into a single investor-owned monopoly.

30. Armstrong, Cowan, and Vickers, *Regulatory Reform,* 255n8. British Gas was afterward an active participant in buying gas businesses in Canada, Argentina, India, Brazil, and elsewhere.

31. That the United Kingdom created problems by its failure to restructure before privatizing British Gas is not particularly controversial among economists. The UK government itself admitted as much when it chose a radical restructuring of its government-owned electricity business in the 1990s prior to that privatization. See Armstrong, Cowan, and Vickers, *Regulatory Reform*, 255.

32. The original Gas Act of 1986 did specify that the director general of the Office of Gas Supply (Ofgas) should "enable persons to compete effectively in the supply of gas through pipes at rates which, in relation to any premises, exceed 25,000 therms a year" (part 1, section 4 [2][d]). However, without explicit technical/informational mechanisms or cooperation from British Gas in providing such open access, little success was achieved in this regard.

33. UK Competition Commission, Office of Fair Trade, "Gas and British Gas plc, Reports under the Gas and Fair Trading Act" (London: Monopolies and Mergers Commission, 1992).

34. The Network Code is a very complex and expensive software management package used to run the entry/exit system. It was introduced in 1996 after two years of negotiations with Ofgas and industry participants. Shippers in the United Kingdom by the late 1990s had been virtually unanimous in condemning the Network Code as too complex, too burdensome, and a barrier to efficient trading. For example, the managing director of Alliance Gas called the Network Code "fundamentally flawed" and declared that the daily balancing regime was "absolutely unnecessary." *UK Gas Report* 115 (Nov. 1996).

35. I was a part of those seminars with customers and shipper-traders, representing British Gas.

36. Another reason that the move initially proposed by British Gas in late 1995 was dropped was the incompatibility of the computer and management systems developed to implement the single NBP regime to handle a more realistic pipeline tariff regime.

37. Title XII of the treaty states: "To help achieve the objectives referred to in Articles 7a and 130a and to enable citizens of the Union, economic operators and regional and local communities to derive full benefit from the setting up of an area without internal frontiers, the Community shall contribute to the establishment and development of trans-European networks in the areas of transport, telecommunications and energy infrastructures." Treaty of European Union, Title XII: Trans-European Networks, Article 129b, Official Journal C 191, July 29, 1992.

38. Directive 98/30/EC of the European Parliament and of the Council.

39. Directive 2003/55/EC of the European Parliament and of the Council, replacing the directive of 1998.

40. "Every common carrier subject to the provisions of this act shall, according to their respective powers, afford all reasonable, proper and equal facilities for the receiving, forwarding and delivering of . . . property . . . and shall

not discriminate in their rates and charges." Interstate Commerce Act of 1887, Section 3.

41. *DG Competition Report on Energy Sector Inquiry*, Brussels, European Commission, Jan. 10, 2007.

42. Ibid., 7.

43. Ibid., initial paragraphs 50, 59.

44. Directive 2009/73/EC of the European Parliament and of the Council of July 13, 2009, concerning common rules for the internal market in natural gas and repealing Directive 2003/55/EC.

45. Proposal for a directive amending Directive 2003/55/EC (presented by the Commission), 2007, initial paragraph 11.

46. There are smaller pipeline systems in South Africa (which has a reasonably extensive government-owned oil and gas pipeline system) and New Zealand (which developed its pipelines at about the same time as Australia).

47. I was involved in a number of pipeline matters in Australia from 1995 through 2001 that resulted, either directly or indirectly, from that 1993 study, representing the state of Victoria, the National Competition Council, BHP (the gas producer in the Bass Strait), and other gas users in New South Wales on various matters involving the pipelines in eastern Australia.

48. Very large gas reserves were discovered in the northwest part of Australia in the 1960s, and they have fueled a large liquefied natural gas export business in the Pacific. A pipeline was also constructed in 1990 from Alice Springs to Darwin in the Northern Territory. The northwest lines, however, are not linked to those in eastern Australia or the country's major population centers.

49. The much smaller gas operations in Brisbane were served by nearby Queensland suppliers.

50. Without a national regulator for gas transmission, and without a strong "commerce clause" like that in the US Constitution, it has been traditional practice in Australia for states to safeguard their own natural resources. Indeed, the inability to safeguard and regulate interstate trade at the Commonwealth level had long resulted in inconsistent policies, overlapping institutions, and inefficient investment in Australia.

51. According to Hawke:

> The Trade Practices Act is our principal legislative weapon to ensure consumers get the best deal from competition. But there are many areas of the Australian economy today that are immune from that Act: some Commonwealth enterprises, State public sector businesses, and significant areas of the private sector, including the professions. . . . This patchwork coverage reflects historical and constitutional factors, not economic efficiencies; it is another important instance of the way we operate as six economies, rather than one. . . . The benefits for the consumer of expanding the scope of the

Trade Practices Act could be immense: potentially lower profes-
sional fees, cheaper road and rail fares, cheaper electricity. (W. J. R.
Hawke, "Building a Competitive Australia" [ministerial statement,
Mar. 12, 1991])

52. *Report by the Independent Committee of Inquiry, National Competition Policy* (Canberra: AGPS, 1993), referred to as the "Hilmer report," after its
chair, Frederick G. Hilmer, then dean and director of the Australian Graduate
School of Management, University of New South Wales.

53. "Government-owned businesses constitute 10% of Australia's GDP [gross
domestic product]. . . . In the case of rail, electricity, water and gas utilities, for
example, the Industry Commission has identified opportunities for increasing
GDP by over 2%, or $8 billion per annum." Hilmer report, 129.

54. Ibid., 130.

55. Ibid., 218–19, 221–22.

56. The enthusiasm was assisted by the prospect of substantial "competition
payments" from the Commonwealth government to states that cooperated with
the reforms.

57. The pipeline serving Sydney also served the federal capital of Canberra
and included a line running down toward, but not reaching, the border with Victoria. The Victorian system served Melbourne and western/northern outlying areas heading toward, but again not reaching, the border with New South Wales.

58. The Australian government sold the Moomba-Sydney natural gas transmission pipeline to Australia Gas and Light (51 percent) and Nova Corp. of Canada and Petronas of Malaysia (49 percent) for A$534 million.

59. Hilmer report, 226–27. It is unreasonable to think that those who drafted
the Hilmer report were unaware of the problems emanating from the 1986 privatization of monopolistic British Gas or the favorable reports of the 1991 privatization of a restructured Gas del Estado in Argentina.

60. Productivity Commission 2001, *Review of the National Access Regime*,
report no. 17 (Canberra: AusInfo, 2001), 42.

61. AGL renamed the line the Eastern Australian Pipeline (EAPL).

62. By "notional," I mean a transport system conceived as a central repos-
itory for gas, in which all gas supplies are assumed to travel to a central hub
where the sale takes place before traveling onward to be delivered to customers.
In this respect, the Victorian system mirrored the conception of the transport
system in the United Kingdom, where all gas is assumed to travel to and from the
"national balancing point."

63. The label "market carriage" was Orwellian. It was a mandatory bidding
regime for pricing delivered gas reflecting short-term and often unpredictable
system operational conditions. It denied shippers the ability to contract for pipeline capacity. VENCorp was later absorbed into the Australian Energy Market
Operator with its peculiar gas "market carriage" mandate intact.

64. The Victorian pool-based gas-trading system was the brainchild of the state's principal external consultants, who saw the opportunity to mimic the role of VENCorp in its electricity Independent System Operator (ISO) role for the gas system. The move was strongly opposed at the time by BHP (the single gas supplier to Victoria), which saw the use of such electricity-market-inspired commercial trading arrangements for gas and pipeline costs as wastefully bureaucratic, inefficient, and generally unnecessary. I represented BHP at the time in public forums in Melbourne regarding the thoughtlessness of the proposed regime. But BHP made no headway against the state in the latter's choice of what it considered to be a convenient, electricity-inspired commercial regime for the privatization of state-owned assets.

65. The direct costs, in annual fees billed by VENCorp, for operating that system have been reported at A\$18–A\$20 million. The indirect costs of the system, which reflect barriers to interstate pipeline shipments for an arrangement that forbids pipeline transport contracts, are surely much higher.

66. AGL Cooper Basin Natural Gas Supply Arrangements, ACompT 2 (Oct. 14, 1997), 9.

67. Ibid., 15.

68. Ibid., 74.

69. Fearing that, if the project were built starting from the basin, foreign companies might try to intervene and change the destination of the pipeline, the government began the construction at the Buenos Aires terminus. By starting there, the government could ensure that the ten-inch pipeline would serve its intended market.

70. International Energy Agency, *World Energy Statistics and Balances* (Paris: OECD, 1990), 210. By 2007, gas accounted for 51 percent of Argentina's primary energy. See Luis A. Erize and Sergio M. Porteiro, "Argentina," *Gas Regulation 2007* (London: Global Legal Group, 2007).

71. The Dutch government eventually assumed Cogasco's \$1.017-billion debt. The Argentine government subsequently paid the debt to the Dutch government and took control of the pipeline.

72. See Ann Davison, Chris Hurst, and Robert Mabro, *Natural Gas: Governments and Oil Companies in the Third World*, Oxford Institute of Energy Studies (Oxford: Oxford University Press, 1988), 109–12.

73. Gas del Estado, *Boletín Estadístico Anual, 1989* (Buenos Aires: Ministerio de Economía, Centro de Documentación, 1989), 16.

74. NERA, *Final Report: Argentina Gas Tariff Study* (White Plains, NY: National Economic Research Associates, 1991), 7. I was involved in the privatization of Gas del Estado in 1991, working for the World Bank and Ministry of Economy to define and structure a tariff regime for independent transport and distribution companies. It was during this undertaking that I again worked with Sir James McKinnon (we had worked together to recommend the restructuring

of the Polish gas system in 1990, also for the World Bank), who was the first gas industry regulator in the United Kingdom for the newly privatized British Gas. McKinnon related to me himself how he discovered, through trial and error, his considerable regulatory powers in spite of that company's obstructive behavior toward regulation generally.

75. British Gas reversed its earlier position and participated in the privatization anyway, purchasing the main distributor in Buenos Aires, which it named MetroGAS.

76. Hafees Shaikh, Manuel A. Abdala, et al., "Argentina Privatization Program: A Review of Five Cases," Case Study 3: Gas del Estado (Washington, DC: World Bank, 1995), 144.

77. "Argentine Gas Sell-Off a Success," *Financial Times*, Dec. 4, 1992, 36.

78. In other words, during the privatization of Gas del Estado, the government of Argentina placed the welfare of its gas consumers first, and the size and/or protected profitability of the privatized businesses (and the direct proceeds from the privatization) second. Such structural separation and promotion of competition did not accompany the subsequent privatization of YPF, the state-owned gas-producing firm, which was privatized in 1993 as a dominant single gas supplier (controlling about 70 percent of gas sales to distributors and other consumers in Argentina).

79. The government did, however, give the two privatized pipelines—to the south and north of Buenos Aires—exclusive franchises. In the United States, the regulator grants no such thing. This was a concession to buyers wishing to be protected from entry in their particular regions—an understandable incentive on a pipeline system built by a government without the competitive pressures from the markets for either gas or capital.

80. Those institutional difficulties are dealt with at length in chapter 6.

81. US Department of Justice, *Competition in the Oil Pipeline Industry: A Preliminary Report* (Washington, DC: Antitrust Division, Department of Justice, May 1984).

82. US Department of Justice, *Oil Pipeline Deregulation* (Washington, DC: Antitrust Division, Department of Justice, May 1986). The DOJ cited two contributions in particular to its final report: John A. Hansen, *U.S. Oil Pipeline Markets* (Cambridge, MA: MIT Press, 1983); and Robert E. Anderson and Richard T. Rapp, *Competition in Oil Pipeline Markets: A Structural Analysis* (White Plains, NY: National Economic Research Associates, 1978). The Hansen study was his PhD dissertation at Yale University. The Anderson and Rapp study was done for "a group of independent refiners that were concerned about the impact of deregulation on oil pipelines rates" (1).

83. DOJ (1986), xv. The 1986 DOJ report has heavily influenced the analysis of pipeline markets in the United States since. It employed the familiar Hirfindahl-Hirschman index (HHI), which is the sum of the squares of the mar-

ket shares, measured on a scale of 0 to 100, within "origin markets" and "destination markets." Employing a rule of thumb that oil cannot be shipped economically outside a seventy-five-mile radius by road, the study used standard US statistical regions to calculate HHIs for the origin and destination market of all major US oil pipelines. For those pipelines with an HHI of less than 2,500 (representing four equally sized lines, or four times 625), the DOJ did not recommend the continuation of standard cost of service regulation. The DOJ continues to use these definitions of competition in origin and destination markets when evaluating prospective mergers that involve pipeline companies.

84. Energy Policy Act of 1992, H.R. 776, Title XIII, Sec. 1801, 1803.

85. This was called Order No. 561-A, FERC Stats. & Regs (Regs Preambles, 1991–1996), at 30,985 (1993).

86. The 1 percent was not based on any particular productivity study for the industry.

87. See 18 CFR Part 342, FERC Order in Docket No. RM05–22-000 (Mar. 21, 2006), and Docket No. RM10–25-000 (Dec. 16, 2010).

88. To the extent that it appears that the real rising price caps for regulated oil pipelines continue to bind (meaning that oil pipeline prices uniformly move upward according to the cap), or that reported oil pipeline profitability rises unduly with the latest decision, the FERC may revisit in the future whether that experience contradicts the 1986 DOJ conclusion that crude oil pipelines possess insufficient market power to justify continued cost-based regulation or whether the continued, generally mechanical, increase in the cap on regulated prices warrants a reexamination of the underlying methodology.

89. DOJ (1986), xvi.

90. Comments of the US Department of Justice in Response to Notice of Technical Conference, Docket No. OR92–6-000, July 30, 1992.

91. Ibid.

92. See Docket No. RM-95–6-000, Order Reversing Initial Decision, 70 FERC ¶ 61,139 (Feb. 8, 1995).

93. "Koch has not met the requirements of the Policy Statement and has not shown that it lacks market power." Docket No. RM-95–6-000, Order Reversing Initial Decision, p. 23.

94. These twin reasons for rejecting gas pipeline deregulation—the small geographic market relevant for gas pipelines and the lack of realistic alternatives even if other pipelines are nearby—was also a factor in the decision by the German pipeline regulator, BNetzA, in separate decisions issued in September and October 2008, to deny exemption from cost-based regulation for all of the pipeline systems operating in Germany. The regulator cited classical market share indexes, the long-term booking of the dominant share of capacity, and the prominence of affiliate transactions as the reasons for rejecting pipeline requests for

exemptions from cost-based regulation. See "Decision to Tighten Gas Grid Competition," press release, Federal Network Agency, Bonn, Oct. 21, 2008.

95. In 2011, the APO group controlled the privatized pipelines in Victoria and New South Wales, including the Interconnect and the pipeline from Victoria to South Australia. Jemena owns the EGP. Epic Energy owns the old pipeline from the Cooper basin to Adelaide and the pipelines supplying Brisbane.

96. The National Competition Council (NCC) was established by all Australian governments in November 1995 to act as a policy advisory body to oversee their implementation of the recommendations of the Hilmer report.

97. Duke Eastern Gas Pipeline Pty Ltd (2001) ACompT 2 (May 4, 2001), paragraphs 114–15. In that case, I was the witness for the NCC on the question of extending regulatory coverage to the EGP, as noted in that decision.

98. From Minister Macfarlane's official bio, his nickname "Chainsaw" reflected his direct approach "to 'cutting through' the red tape to get things done for Australian industry on a political level."

Chapter 5

1. Of course, pipelines that take gas away from liquefied natural gas (LNG) terminals or from oil import terminals have no such built-in obsolescence, but they still face a dynamic and changing marketplace.

2. The Panhandle Eastern Pipe Line Company constructed a twelve hundred–mile pipeline route from the Kansas/Oklahoma region of the United States to the markets in Chicago, Detroit, Indianapolis, and farther east in 1931. The line came within thirty-five miles of the major metropolitan area of Kansas City without connecting to the city. The reason was that the owner of the local gas company, Henry Dougherty, preferred to protect his monopoly position in Kansas City by supplying gas through his own affiliated company, Cities Service Company. See Castaneda and Smith, *Gas Pipelines*, 15–49. It took a bold and politically savvy pipeline developer, Dennis Langley, of Kansas Pipeline (once counsel to the US Senate Judiciary Committee and later chairman of the Kansas Democratic Party), to overcome the lock that Cities Service (later Williams Companies) had on Kansas City and connect the city to the Panhandle System in the 1990s over the strenuous entry-deterring tactics of Williams. I was a witness for Kansas Pipeline in the various civil and administrative court actions associated with its ultimately successful entry into Kansas City.

3. In early 1942, U-boats were sinking up to a dozen oil tankers a month on the East Coast of the United States. It was an extreme national emergency, for not enough oil reached the cities of Washington, New York, and Boston to fuel the rapidly expanding war effort. From a tiny share of oil deliveries in 1941, two

new oil pipelines turned the tide in East Coast oil shipments to become the biggest source of East Coast oil by the end of the war. They were the first large-diameter, long-distance petroleum pipelines in the world. Called "Big Inch" and "Little Big Inch," they pushed both the legal and technological envelope for pipelines. But at the end of the war, the oil industry chiefs, who generally were highly mistrustful of one another, could not get rid of this cooperative project fast enough. Thus, the two pipelines that so contributed to the war effort became wartime surplus. There was no unanimity on what do to with them. *Time* magazine of August 12, 1946, wrote: "One jokesmith wanted the lines for piping grapefruit juice from Texas to New York. Another thought tough Texas jackrabbits could be profitably run to eastern markets since 'anything becomes a delicacy if it is moved far enough.' Even harried WAA [War Assets Administration] officials took time out to join in the fun. Their proposal: start carbonated water through the pipes in Texas, spike it with bourbon in Kentucky, route the piped highballs through the 'ice mines of Appalachia' and then on to the bars of Manhattan." Eventually, the newly formed Texas Eastern Gas Pipeline Company won the bid to acquire the assets and convert them to natural gas.

4. See C. Ménard and M. M. Shirley, eds., *Handbook of New Institutional Economics* (Dordrecht, Netherlands: Springer, 2005); and Furubotn and Richter, *Institutions and Economic Theory*.

5. Douglass C. North, "Economic Performance through Time," *American Economic Review* 84, no. 3 (June 1994): 360.

6. Ibid., 359.

7. For transaction cost generally, see Steven Tadelis, "Complexity, Flexibility, and the Make-or-Buy Decision," *American Economic Review* 92, no. 2 (May 2002); Patrick Bajari and Steven Tadelis, "Incentives versus Transaction Costs: A Theory of Procurement Contracts," *RAND Journal of Economics* 32, no. 3 (Autumn 2001); and George Baker, Robert Gibbons, and Kevin J. Murphy, "Relational Contracts and the Theory of the Firm," *Quarterly Journal of Economics* 117, no. 1 (Feb. 2002). A more formal theory of property rights has been developed by Oliver Hart, John Moore, and Sanford Grossman. See Hart and Moore, "Property Rights and the Nature of the Firm," *Journal of Political Economy* 98, no. 6 (1990); Grossman and Hart, "The Costs and Benefits of Ownership: A Theory of Vertical and Lateral Integration," *Journal of Political Economy* 94, no. 4 (1986); and Hart, "Corporate Governance: Some Theory and Implications," *Economic Journal* 105, no. 430 (May 1995).

8. There is continuing debate between those who consider the field of new institutional economics to be either "less a new body of theory or methods than a change in emphasis" or much the same thing as the field called "law and economics," as suggested by Judge Richard Posner. "The New Institutional Economics Meets Law and Economics," *Journal of Institutional and Theoretical Economics* 149, no. 1 (1993): 73–87. Coase's reply to Posner was that the new institutional

economics seeks to do away with "the kind of [microeconomic] abstraction which does not help us to understand the working of the economic system." "Coase on Posner on Coase," *Journal of Institutional and Theoretical Economics* 149, no. 1 (1993): 96–98. From the perspective of the present industry study, Coase—not Posner—emerges as the clearer and more insightful voice.

9. The contrast between the benefits of analyzing "specific phenomena" of transaction cost economics and "mathematical theory" of agency theory is highlighted by Scott Masten and Stéphane Saussier: "Agency theorists, with their emphasis on axiomatic deduction, have been hesitant to incorporate into their models constraints, such as bounds on cognitive ability, that cannot be easily modeled. Transaction-cost economists working in the tradition of Ronald Coase and Oliver Williamson, by contrast, have sought to develop and refine theory guided by more specific phenomena or puzzles than by the susceptibility of the theory to mathematical modeling." "Econometrics of Contracts: An Assessment of Developments in the Empirical Literature of Contracting," in *Economics of Contracts: Theories and Applications*, ed. Eric Brousseau and Jean-Michel Glachant (Cambridge: Cambridge University Press, 2002), 288–89. As this book shows, it is only through such a detailed analysis of institutional foundations of the two closely related oil and gas pipeline businesses that the reasons for their current differences become apparent.

10. The contribution of the SEC to a revolution of the industry of business information is well described in Thomas K. McCraw, "Landis and the Statecraft at the SEC," chap. 5 in *Prophets of Regulation* (Cambridge, MA: Harvard University Press, 1984), 153–209.

11. In the language of transaction cost economics, this low-information environment exacerbates the bounded rationality problem in writing complete contracts—leading to a greater incentive toward vertical integration.

12. For a review of the body of literature on these elements that lead firms like pipelines to integrate vertically, see Paul Joskow's 2005 survey article, "Vertical Integration."

13. Benjamin Klein, Robert G. Crawford, and Armen A. Alchian describe quasi rents and their potential appropriation using an example of a capital investment in a printing press business contracted to a publisher. Without another potential publisher customer around, the quasi rent on the installed machine is equal to the amortized fixed cost of the press minus its salvageable value. "Vertical Integration, Appropriable Rents, and the Competitive Contracting Process," *Journal of Law and Economics* 21, no. 2 (1978): 298–99.

14. Williamson, *Mechanisms of Governance*, 377–78.

15. Klein, Crawford, and Alchian, "Vertical Integration," 298.

16. In modern fuel markets, one may ask whether contracting is still as costly as it was when the pipeline transport industry was young. The decades-long durability of pipeline assets coupled with the volatility of fuel markets—and the

opportunism of parties, as Williamson would define it—almost guarantees substantial ex post contracting costs.

17. Oliver E. Williamson, "Transaction-Cost Economics: The Governance of Contractual Relations," *Journal of Law and Economics* 22, no. 2 (1979): 247.

18. According to Williamson, there are three generally recognized categories of such specificity: (1) site specificity, where parties to a transaction are in a "cheek-by-jowl" relationship tied to a particular location; (2) physical asset specificity, where the investments made in equipment and machinery pertaining to that specific transaction have low or no value in alternative use; and (3) dedicated, long-lived assets, where parties make investments that would not otherwise be made but for the prospect of selling a significant amount of product to a particular customer over a period of time. See Oliver E. Williamson, "Credible Commitments: Using Hostages to Support Exchange," *American Economic Review* 83, no. 4 (Sept. 1983): 526.

19. Consider again the example from Klein of the printing press and publisher. The extent to which publishers could be held up by printers, and hence wish to own rather than contract for printing facilities, is highly dependent on whether what is printed is time-sensitive or tied to a particular location. For example, newspaper publishers generally own their presses, while book publishers do not. For the newspaper publisher with papers to ship on a rigid local schedule, the inability to secure a local press would expose appropriable quasi rents to seizure by a printer. Book publishers, however, have considerably more flexibility with respect to both timing and location. They are not tied to any particular region or publisher and therefore do not so expose themselves to appropriable quasi rents.

20. See J. Harold Mulherin, "Complexity in Long-Term Contracts: An Analysis of Natural Gas Contractual Provisions," *Journal of Law, Economics, and Organization* 2, no. 1 (Spring 1986): 105–17.

21. Klein, Crawford, and Alchian, "Vertical Integration," 310–11. Why are the owners of oil wells and refineries more at risk of opportunistic behavior by the pipeline owner than the other way around? The example of Klein and his colleagues assumes competitive production and refining sectors connected by a single pipeline. If the tables were turned and there were more than one transport route (another pipeline or a river suitable for oil barges, for example) and concentrated production and refining sectors, then it would be the pipeline, not the producers/refiners, who would be subject to quasi-rent appropriation.

22. As Stéphane Saussier describes, "When economic agents decide to collaborate, they usually create a 'contract interface' to guide the transaction, the subject of the collaboration. To maximize the gains the interface must be correctly designed." "Transaction Costs and Contractual Incompleteness: The Case of Électricité de France," *Journal of Economic Behavior* 42 (2000): 190.

23. The discussion of the origin of common carriage in this section draws on Daggett, *Principles of Inland Transportation*, 284–335; and Keeler, *Railroads, Freight and Public Policy*, 19–42.

24. An important element of such grants was the funding of the local sovereign or other governmental authority. The modern grant of licenses and franchises, with the various taxes, fees, transfers, and so forth, that accompany them, is a close analogue to the medieval practice. Public utilities and other regulated enterprises continue to make reliable, albeit indirect and opaque, mechanisms to raise government funds in lieu of more direct or transparent taxes.

25. Dudley F. Pegrum, "Restructuring the Transport System," in *The Future of American Transportation*, ed. Ernest W. Williams Jr. (Englewood Cliffs, NJ: Prentice-Hall, 1971), 63.

26. Daggett, *Principles of Inland Transportation*, 301.

27. Of course, proscribed practices under the common law, in the absence of specific statutes, merely invite plaintiff lawsuits to remedy an injustice. The question of whether such lawsuits would be adequate to control more rampant or complex abuses, like those involving Standard Oil, was very much on the mind of President Theodore Roosevelt when, in 1906, he recommended that the Senate take prompt legislative action to impose federal pipeline industry regulation instead.

28. Pegrum, "Restructuring the Transport System," 63.

29. There are many stories in the early history of oil pipeline transportation, before the certificate process, in which new pipeline developers raced to keep ahead of competitive railroad interests in securing rights-of-way. Pipeline developers secured rights through third parties, communicated in cipher, and availed themselves of circuitous routes as railroads and other competitors sought to disrupt their plans to ship oil to market. See Johnson, *Development of American Petroleum Pipelines*, chaps. 2 and 3.

30. Davis and North, *Institutional Change and American Economic Growth*, chap. 7 (135–66). Davis and North drew for their analysis on Paul MacAvoy, *The Economic Effects of Regulation* (Cambridge, MA: MIT Press, 1965).

31. Interstate Commerce Act of 1887, 24 Stat., p. 380. The use of the 1887 act as a vehicle to promote cartel behavior and prohibit price cutting was further strengthened when the railroads pushed through Congress the Elkins Act of 1903, which called for treble damages for instances of price cutting. The story will return to the Elkins Act in chapter 6 when Assistant Attorney General Thurman Arnold used its provision to attack vertically integrated oil pipelines in 1941.

32. Davis and North, *Institutional Change and American Economic Growth*, 51.

33. The result of that collapse was the Staggers Act, signed by President Jimmy Carter in 1980, which did away with the common carriage restrictions of

the then-century-old Interstate Commerce Act. See Keeler, *Railroads, Freight and Public Policy*, 97–114.

34. Various "hybrid" forms of governance are outlined in Oliver E. Williamson, "Comparative Economic Organization: The Analysis of Discrete Structural Alternatives," *Administration Science Quarterly* 36, no. 2 (1991): 269–96. Claude Ménard notes that such hybrids can take a variety of forms suited to various purposes. "The Economics of Hybrid Organizations," *Journal of Institutional and Theoretical Economics* 160, no. 3 (2004): 160.

35. See Davis and North, *Institutional Change and American Economic Growth*, chap. 7 (135–66).

36. As Avinash Dixit has pointed out, the word *governance* appears only five times in the literature in the 1970s but more than thirty thousand times by the end of 2008. Reflecting this shift in emphasis, Dixit explicitly enumerated the themes of (1) security of property rights, (2) enforcement of contracts, and (3) collective action as the bedrock of economic activity. "Governance Institutions and Economic Activity," 5.

37. John R. Commons, *Legal Foundations of Capitalism* (New York: Macmillan, 1924), 7. This volume is remembered more by legal than economic scholars, as evidenced by its being the only one of Commons's five books still in publication—by a publisher of legal reference texts.

38. I defer the specific legal references until the next two chapters, which deal with these court cases in greater detail.

39. Commons provided an economic discussion of the circumstances under which, in 1890, the US Supreme Court first recognized the existence of "intangible" as opposed to "corporeal" property. See Commons, *Institutional Economics*, 649–56.

40. An extended discussion of the contrast between the common-law and civil law version of property rights is contained in Furubotn and Richter, *Institutions and Economic Theory*, 76–85.

41. In America, we think concretely according to the common law method of individual cases and precedents, conformable to our judicial sovereignty; while the Europeans think abstractly in deductive terms handed down from Justinian, Napoleon, Adam Smith and Ricardo. If we generalize . . . we discuss only general principles, leaving their application to investigations of the particular cases. In this way has arisen the American common law method. This American system of custom, precedent, and assumptions is with difficulty comprehended by European economists and jurists who operate under a system of codes constructed originally . . . on the model of the perfected Roman law and changeable only by legislatures. It is even understood with difficulty by the British, whose legislature is superior to the judiciary. (Commons, *Institutional Economics*, 713)

This was later restated by Kenneth Parsons in Commons, *Economics of Collective Action*, app. 3, 341.

42. Of course, the "superiority" of the Supreme Court is still constrained by the US Constitution's checks and balances. The justices themselves are appointed by the president and confirmed by the Senate. The Constitution itself can be amended by three-fourths of the legislatures of the fifty states. And the Supreme Court itself may come to a new view of the facts of a particular kind of case and change its mind. These checks, balances, and constraints notwithstanding, the Supreme Court at any particular time is the final authority in the United States of what is constitutional and what is not.

43. Romanic ownership can be thought of as a box, with the word "ownership" written on it. Whoever has the box is the "owner." In the case of complete, unencumbered ownership, the box contains certain rights, including that of use and occupancy, that to the fruits or income, and the power of alienation. The owner can, however, open the box and remove one or more such rights and transfer them to others. But as long as he keeps the box, he still has the ownership, even if the box is empty. The contrast with the Anglo-American law of property is simple. There is no box. There are simply various sets of legal interests. (John H. Merryman, "Ownership and Estate [Variations on a Theme by Lawson]," *Tulane Law Review* 48 [June 1974]: 927)

44. See Gillian K. Hadfield, "The Many Legal Institutions That Support Contractual Commitments," in *Handbook of New Institutional Economics*, ed. C. Ménard and M. M. Shirley, 175–203 (Dordrecht, Netherlands: Springer, 2005); and Rafiel La Porta, Florencio Lopez-de-Silanes, Andrei Shleifer, and Robert W. Visny, "Law and Finance," *Journal of Political Economy* 106, no. 6 (Dec. 1998): 1113–55. The status of long-term energy contracts in Europe remains unsettled in the early twenty-first century. Among other problems are cases in which a state-owned company either granted special, low tariffs to certain categories of users or bought energy from producers at prices above market rates. In such situations, contract holders have an untested claim to state compensation. *Platts EU Energy*, "Long-Term Contracts: A Legal Quagmire," no. 174 (Jan. 11, 2008), McGraw Hill, 8–9.

45. When Victoria wished to create a special gas pool arrangement for its privatized pipeline system in 1998, it asked for and received, without apparent dissent, a special exception to the *Australian National Third Party Access Code for National Gas Pipelines* to accommodate what Victoria called "market carriage" (see section 3.7, "Capacity Management Policy").

46. A celebrated example is John R. Commons's invention of workmen's compensation insurance. He wrote novel legislation in 1911 to give employers (and their private insurers) a financial stake in worker safety. Up to that time,

the only common-law remedy for worker injury was to sue the foreman, and the only thing the government did to try to help was to provide hated "safety police." Workmen's compensation eliminated the safety police (the best of whom were then hired as risk experts for Wausau Mutual Insurance at enhanced salaries reflecting the profits their talents could earn for Wausau); accident rates dropped precipitously; and the workmen's compensation insurance premiums, which the legislation obliged employers to pay, were far more than repaid in greater productivity. See Commons, *Economics of Collective Action*, 279–84.

47. Richard W. Hooley, *Financing the Natural Gas Industry* (New York: AMS Press, 1968), 13, 45, 50. The insurers include New York Life Insurance Company, Teachers Insurance and Annuity Association, and Phoenix Mutual Life Insurance Company.

48. Among them are Professors Emory Troxel, of Wayne University; Walter Splawn, of the University of Texas; James Bonbright, of Columbia University; Law Professor Eugene Rostow, of Yale University (who first investigated the principles of pipeline contract carriage in 1952); Martin Glaeser, of the University of Wisconsin; and Eli Clemens, of the University of Maryland—to say nothing of John R. Commons, who began the scholarly study of the economics of utility price regulation.

49. An example occurred in 1995, when British Gas's regulator, Ofgas, published a draft series of proposals for a five-year price-control period for BG that effectively abandoned a 1993 decision by the UK Monopolies and Mergers Commission on calculating permissible revenue. The decision effectively removed approximately three billion pounds from the company's asset base. As a result, the stock price of British Gas fell approximately 24 percent in two days and its debt securities were downgraded three steps by Standard & Poor's, the debt-rating agency.

50. Jack Stark, "The Wisconsin Idea: The University's Service to the State" (Madison: Legislative Reference Bureau, 1995), 17. Reprinted in *Wisconsin Blue Book, 1995–1996* (Madison: State Printing Office, 1995).

51. As Olson much later said about Commons, "The basis of Commons' thinking was the view that the market mechanisms did not of themselves bring about fair results to the different groups in the economy, and the conviction that this unfairness was due to disparities in the bargaining power of these different groups." Olson, *Logic of Collective Action*, 115. Commons sometimes took this point of view to the extreme, suggesting, for example, that the direct election of representatives of each pressure group would form an effective legislature for the country.

52. Commons, *Economics of Collective Action*, 2.

53. "Characteristically, the establishment and enforcement of institutional norms requires some kind of *collective* action, private or public. This is precisely what the new institutional economics analyses, and it is this recognition of the

need for collective action that sets the NIE apart from orthodox neoclassical theory." Furubotn and Richter, *Institutions and Economic Theory*, 20.

54. Commons, *Institutional Economics*, 712.

55. "The proposition that economic organization has the purpose of promoting the continuity of relationships by devising specialized governance structures, rather than permitting relationships to fracture under the hammer of unassisted market contracting, was thus an insight that could be gleaned from Commons." Williamson, *Economic Institutions of Capitalism*, 3. Commons's legacy was somewhat controversial. Williamson went on to say that "not everyone associated with the lens of contract would agree. Coase, for example, contends that 'American institutionalism,' of which Commons was a prominent part, 'is a dreary subject. . . . All it had was a stance of hostility to the standard economic theory. It certainly led to nothing.' . . . My view is that Commons was ahead of his time. He had a lens of contract conception of economics as early as the 1920s." Williamson, "Transaction Cost Economics," 43.

56. Geoffrey Hodgson, "John R. Commons and the Foundations of Institutional Economics," *Journal of Economic Issues* 37 (Sept. 2003): 570.

57. Professor Martin Glaeser, who was as an undergraduate a member of Commons's first class in regulatory economics in 1906, did his graduate work at Harvard under Roscoe Pound and returned to teach at Wisconsin from 1919 until his retirement in 1959, writing the first public utilities text in 1927. Glaeser, *Outlines of Public Utility Economics*. He trained a generation of US regulatory economists. See Harry M. Trebing, "Martin G. Glaeser," in *Pioneers of Industrial Organization*, ed. Henry W. de Jong and William G. Shepherd (Cheltenham, UK: Edward Elgar, 2007), 190–93.

58. Olson, *Logic of Collective Action*, 23–27.

59. Olson, "Collective Action," p. 3 of 5.

60. See Torsten Persson and Guido Tabellini, *Political Economics: Explaining Economic Policy* (Cambridge MA: MIT Press, 2000). Much of this literature references the model of legislative behavior presented by David P. Baron and John A. Ferejohn, "Bargaining in Legislatures," *American Political Science Review* 83, no. 4 (1989): 1181–1205. A recent survey of much of the literature appears in Spiller and Liao, "Buy, Lobby or Sue."

61. Mancur Olson, *The Rise and Decline of Nations: Economic Growth, Stagflation, and Social Rigidities* (New Haven, CT: Yale University Press, 1982), 44.

Chapter 6

1. Most economists would agree that the ICC did no better with respect to the railroads, which dominated the agency's attention from the start. Rather than

promote efficient operation of the rail system in the United States, the ICC is widely credited with crippling that industry by the 1970s.

2. There remains little controversy over the intent of Standard Oil to bar entry through its dealing with railroads. But there may still be some. As Kahn said in 2007, "Only the economically brainwashed can deny that price discrimination has also been used as a means of predation to the ultimate injury of consumers, however frequent routine allusions to [the] proffered—and later refuted— demolition of the contentions of the populists about the tactics used by John D. Rockefeller." Alfred E. Kahn, "Telecommunications: The Transition from Regulation to Antitrust," *Journal of Telecommunications and High Technology Law* 5, no. 1 (2007): 171. See also John S. McGee, "Predatory Price Cutting: The Standard Oil (N.J.) Case," *Journal of Law and Economics* 289 (1958): 137–69; John S. McGee, "Predatory Price Cutting Revisited," *Journal of Law and Economics* 289 (1980): 289–330; and James A. Dalton and Louis Esposito, "Predatory Price Cutting and Standard Oil: A Re-examination of the Trial Record," *Research in Law and Economics* 22 (2007): 155–205.

3. The Standard [Oil Company] claims that the location of its refiners and the use of pipe lines are natural advantages to which it is justly entitled by reason of the energy and foresight of its managers. While in a measure that is true, it must not be forgotten that these advantages were in part obtained by means of unfair competitive methods during years of fierce industrial strife. . . . The development of the pipe-line system by the Standard Oil Company was the result of special agreements with railroad companies. Furthermore, those so-called natural advantages have been and are being greatly increased by discrimination in freight rates, both published and secret, interstate and State, which give the Standard monopolistic control in the greater portion of the country, and which so limit competition as to practically prevent the extension of the business of any independent to a point which even remotely endangers the supremacy of the Standard. (*Report of the Commissioner of Corporations on the Transportation of Petroleum*, xx)

The Commissioner of Corporations was the precursor to the Federal Trade Commission.

4. Cong. Rec., 59th Cong., 1st Sess., S6358 (May 4, 1906).

5. Ibid.

6. The bill took its name from Representative William P. Hepburn (R-Iowa).

7. Foraker's argument was not based solely in legal or legislative principles. At the same time he was participating in the Senate debate, the gas company in Cincinnati (Foraker's political base) was attempting to secure gas from West Virginia to displace manufactured gas for city lighting purposes and to supply Cincinnati industry. The three hundred–mile gas pipeline project was said to cost

five million dollars, and the gas company in Cincinnati was attempting to secure financing at the very moment when the Senate was debating the Hepburn bill.

8. Cong. Rec., 59th Cong., 1st Sess., S6362 (May 4, 1906).

9. Let me answer the Senator from North Dakota [McCumber] in this way: Take West Virginia, whence, I believe, comes the gas in the case mentioned by the Senator from Ohio [Foraker]. Of course that gas is not going down yonder three or four hundred miles without pressure, and therefore it comes from the surrounding country where originally tapped, and the gas company which has bored its well may be robbing the landowners around it of their gas, because it has great capital and the facilities for sending it off and selling it, while the landowner, who is not able to pipe it to the market, is not allowed to pipe it through the established pipe line, because Congress will not help him. (ibid.)

10. "According to the Senator's [Tillman's] suggestion, we would make [gas pipelines] common carriers by compelling the company that owns the trunk line which has been built, we will say, from Cincinnati to the natural gas fields of West Virginia, to buy all the gas brought to it in West Virginia and transport it, in order that the men there should have the benefit [of the pipeline] . . . without expending the $5,000,000 necessary to make the [pipeline] available" (ibid).

11. His key point was the following:

Nobody is interested in that enterprise, except only the people who are building the line with the idea of bringing the gas to Cincinnati [and substitute it there for manufactured gas], to do a great public service, and they have had trouble enough to set the enterprise on foot. They are just now in the midst of their trouble, trying to raise the money. They have not yet been able to raise it all. If it should go out, after they have raised the money to build the line, that any man can take possession of it to bring gas there for his own purposes, and that the line is to be under the charge of the Interstate Commerce Commission, I think it will be the end of the enterprise. (Cong. Rec., 59th Cong., 1st Sess., S6371 [May 4, 1906])

12. Ibid.

13. Ibid.

14. There is irony in Senator Foraker's role as the champion of modern competitive gas pipeline markets. Elected to the Senate in 1896, he was indeed the only Senate Republican (Roosevelt's party) to vote against the full Hepburn bill, a position that may have been related to payment he received from the Standard Oil Company for legal advice he provided during his first term. When news of this involvement became public in 1908, exposing a seeming conflict of interest, Foraker was forced to retire from Congress. It took three more years to build Foraker's gas pipeline to Cincinnati—a 185-mile, twenty-inch line, completed

in 1909 as a combined venture of the Cincinnati Gas Transportation Company, the Union Gas & Electric Company, and the Columbia Gas & Electric Company, which controlled the line and ultimately absorbed the distributor as part of its extensive holding company. See Walter C. Beckjord, *"The Queen City of the West"—During 110 Years! A Century and 10 Years of Service by the Cincinnati Gas & Electric Company, 1841–1951* (New York: Newcomen Society in North America, 1951), 19.

15. Cong. Rec., 59th Cong., 1st Sess., S6456 (May 7, 1906). The commodities clause was originally presented to the Senate as follows: "It shall be unlawful for any common carrier to transport . . . any article or commodity . . . which may be owned by it or in which it has any interest, excepting such as are necessary for its own use in its business as a carrier and not intended for sale" (ibid., S6461).

16. "Whether the Standard Oil Company or the pipe lines which it owns is a common carrier or not, unless you divorce production from transportation, the [Hepburn] amendment is of no practical value. They are immune from regulation because they are transporting their own goods, and if not so immune, what is the use of attempting to regulate the charges which they shall make for transporting their own products? How can you reach the evil?" Cong. Rec., 59th Cong., 1st Sess., S9108 (June 25, 1906).

17. Cong. Rec., 59th Cong., 1st Sess., S9252 (June 26, 1906). Representative Oscar Gillespie (D-Texas) summed up the general mood in both houses of Congress on the matter of oil pipelines and the commodities clause in a short speech that perhaps typifies better than any other how Congress formed its oil pipeline legislation: "Mr. Speaker, I want to record my dissent to the proposition that in divorcing the carrying business from the ownership of products carried by the carrier that we should make an exception of oil pipe lines. We should make no such exception, in my opinion. But, Mr. Speaker, . . . I believe this report is about the best compromise of all differences that could be reached, and therefore I shall vote for it. [Applause]." Cong. Rec., 59th Cong., 1st Sess., H9584 (June 28, 1906). With that, the House joined the Senate in approving a compromise bill that exempted oil pipelines from the commodities clause.

18. What follows is not a history as such of the industry after 1906—a superb history already exists in Arthur Johnson's *Petroleum Pipelines and Public Policy*. This section, rather, looks at the post–Hepburn Amendment US oil pipeline industry through the lens of transaction cost economics—specifically, how the industry transacted to deal with the incompatibility of common carriage and relationship-specific pipelines. For its historical references, unless otherwise noted, the discussion in this section draws upon Johnson, 69–77, 97–98, 145–50, 199–206, 217, 269, 286–87, and 367–68.

19. The Mann-Elkins Act of 1910 extended the jurisdiction of the ICC in a number of areas. Pipelines were required to designate a Washington, DC, representative to whom notices could be served. Mann-Elkins also authorized the

ICC to undertake rate investigations on its own initiative, leaving no doubt that Section 13 of the Interstate Commerce Act allowed it to do so. See Frank H. Dixon, "The Mann-Elkins Act, Amending the Act to Regulate Commerce," *Quarterly Journal of Economics* 24, no. 4 (Aug. 1910): 593–633.

20. The Mann-Elkins Act created the US Commerce Court to deal with streamlining legal procedures. The Commerce Court handled appeals from the ICC to reduce the burden on federal district courts given the specialized issues arising out of ICC matters.

21. Johnson, *Petroleum Pipelines and Public Policy*, 77.

22. Availing itself of its monopoly of the means of transportation the Standard Oil Company refused through its subordinates to carry any oil unless the same was sold to it or to them and through them to it on terms more or less dictated by itself. In this way it made itself master of the fields without the necessity of owning them and carried across half the continent a great subject of international commerce coming from many owners but, by the duress of which the Standard Oil Company was master, carrying it all as its own. (234 U.S. 548 [1914], 559)

23. Splawn, *Report on Pipe Lines*, 1:62.

24. For all practical purposes, *divorcement* referred to the commodities clause, requiring that independent pipelines not be affiliated with shippers or own the oil they transported.

25. Wolbert, *U.S. Oil Pipe Lines*, 15.

26. Splawn, *Report on Pipe Lines*, 1:lxxviii–lxxix. Splawn also recommended the federal regulation of gas pipelines in his summary.

27. The American Petroleum Institute was the oil pipeline trade association to which the ICC's Bureau of Valuations had turned for help in its investigation of pipeline rate-making valuations in 1934. The ICC's Bureau of Valuations, without ever specifying a uniform accounting system, worked for more than a decade to select "typical pipeline systems" to develop inventory forms, surveys of pipeline companies, and calculations of reproduction and replacement cost figures. Aided, in a sense, by the API, oil pipeline accounting under the ICC would remain vague.

28. As Johnson points out, the prevailing opinion was that Arnold included the Elkins cases for the sake of leverage on the main antitrust question, not for actual prosecution in the courts. *Petroleum Pipelines and Public Policy*, 291. Arnold was a very important figure who revitalized American antitrust enforcement during President Franklin Roosevelt's New Deal. My late colleague Alfred Kahn related to me how, as a young assistant in the Justice Department, he had the opportunity to meet the great Thurman Arnold in 1940—at the height of the latter's battle with the oil companies—to get approval to testify before a congressional committee. Arnold simply asked Kahn whether, as such a young-

ster (Kahn was then twenty-three), he was willing and prepared to testify before Congress. While Kahn remembers answering positively, his most vivid memory of that encounter was that Arnold conducted the entire short meeting on his feet and without his trousers (which he had apparently sent out to be pressed).

29. The whole of the consent decree appears in Wolbert, *American Pipe Lines*, app., 165–69. The decree was signed by twenty major oil companies, twenty-two pipeline companies, and seven affiliates or subsidiaries of major pipeline companies. Shortly afterward, in 1943, Assistant Attorney General Arnold, the chief architect of the Justice Department's cases against the oil pipelines, was appointed to the federal court bench by the Roosevelt administration. With Arnold gone, the Justice Department lost its zeal in pursuing the oil pipelines. The consent decree was, and still is, perceived as having been a hasty affair.

30. He continued that pipelines were "the indispensable link (except for [on] shore and offshore wells) between the oil well and refinery. . . . Control of transportation facilities enables the majors to locate their refineries in the market areas, and has remitted non-integrated refiners largely to shifting and unsatisfactory locations in the field." Rostow, *National Policy for the Oil Industry*, 57–58.

31. Ibid., 58.

32. "The [ICC] and the courts would have to consider, as a question of fact and of law, whether the rate system could be justified on the ground that spot and contract shipments would not involve the provision of 'a like and contemporaneous service,' and would not be made 'under substantially similar circumstances and conditions.'" Rostow and Sachs, "Entry into the Oil Refining Business," 909.

33. Wolbert would finish his career in the 1970s as vice president and general counsel of the Shell Oil Company.

34. Wolbert, *American Pipe Lines*, 10–12.

35. See Cookenboo, *Crude Oil Pipe Lines*. His book, which constituted the publication of his doctoral dissertation at the Massachusetts Institute of Technology, reviewed both previous books and gave his own summation. He held that independent refiners should have access to crude oil "without dependence on the transportation facilities of integrated companies," and that "this can best be accomplished by a compulsory joint-venture system embracing all companies, large or small, which are willing to pay a share of the initial outlay" (167–68). Cookenboo studied under Morris Adelman at MIT, who will return in this story as the principal antagonist of Alfred Kahn from Cornell University as the oil companies battled gas distribution companies over the issue of wellhead gas price decontrol.

36. Indeed, de Chazeau and Kahn dismissed Rostow's suggestions as inconsistent with the restrictions of common carriage and thus themselves begged the question: "The economics of pipeline and refinery construction and operation requires that the pipeline be effectively integrated into the refining and distribu-

tion plans of the oil company." De Chazeau and Kahn, *Integration and Competition in the Petroleum Industry*, 345. In an accompanying note, they observed, "Contractual integration, through long-term commitments between a non-integrated pipeline operator and refiners or suppliers of crude (as urged, for example, by Rostow and Sachs . . .) is definitely inferior. Indeed, it is dubious if a common carrier can bind itself contractually to give specified service to one customer in possible preference to other potential future shippers with all of whom it is required to deal equitably" (ibid., n47).

37. From a perspective thirty years later, commenting on the ease of evading the cap under a still-binding consent decree, the DOJ said the following in testimony before the pipelines' new regulator, the FERC: "Unfortunately, the dividend limitations are set at 7% of overall valuation regardless of the debt-equity capital structure of the pipeline. In response to this situation, debt financing in the industry has risen sharply since 1941. . . . [In 1978], heavy debt financing with minimal equity contributions is commonplace, with debt equity ratios of 90:10 or higher. This leveraging greatly increases the return to total capital." Statement of the Department of Justice, presented by Donald A. Kaplan, chief, Energy Section, Antitrust Division, FERC Docket No. RM-78-2, Valuation of Common Carrier Pipelines (Oct. 23, 1978), 16–17. Hereafter, "Kaplan (1978)."

38. See Blaise Ganguin, *Fundamentals of Corporate Credit Analysis* (New York: McGraw Hill, 2005), 178.

39. Parent guarantees of oil pipeline company debt continue to this day as a custom for US oil pipelines, although with a more transparent cost-of-service rate-making formula under the FERC, the practice no longer serves simply to evade the consent decree cap.

40. Wolbert, *U.S. Oil Pipe Lines*, 243. Wolbert discusses "hell or high water" elements to such contracts. Under such obligations, "if for *any* reason whatsoever, even if the line is inoperable, or the inability to ship is due to causes which under normal commercial dealing would provide a *force majeure* escape, the pipeline does not have sufficient cash on hand to pay the principal and interest on the debt and discharge all its other obligations, the shipper-owners are required to make up the difference by a cash 'deficiency payment'" (ibid., 243–44).

41. For example, a pipeline may classify a peak season in its tariff. In order to secure a significant proportion of prorated capacity during the peak season, the shippers must maintain a level of shipments in the off-peak seasons, so that a twelve-month average can be used as the basis for calculating prorated percentages during the peak. This type of policy does not necessarily favor affiliated shippers, but it does favor consistent shippers.

42. See Johnson, *Petroleum Pipelines and Public Policy*, app. A.

43. In the early years of the twenty-first century, many oil pipelines in the United States have converted from traditional joint ventures into master limited partnerships (MLPs), limited partnerships (LPs), and limited liability com-

panies (LLCs), which have tax advantages under US tax law. See Christopher J. Barr, "Growing Pains: FERC's Reponses to Challenges to the Development of Oil Pipeline Infrastructure," *Energy Law Journal* 28, no. 1 (2007): 61–64.

44. While the ICC never asserted authority over pipeline accounting practices, whatever valuations the ICC could create became the focus of the consent decree rates. The ICC created those valuations by the mid-1940s, but its methods were hidden from the industry and from oil pipeline users for more than thirty years. It was only in 1978 that Jesse Oak, a valuation engineer for the ICC's Bureau of Accounts, revealed in a proceeding before the FERC how the ICC had performed those valuations. That "Oak method" was subsequently criticized heavily by the Justice Department as subjective, circular, and internally inconsistent. See Kaplan (1978).

45. Anderson and Rapp, *Competition in Oil Pipeline Markets*, 2.

46. Kaplan (1978), 9.

47. After considerable litigation, the FERC threw out the ICC's obscure "Oak method" for determining rate base and substituted "trended original cost" (a variant of the commission's gas pipeline rate base method) in 1985. 31 FERC 61,377 at 61,832 (1985).

48. A considerable body of opinion existed at the time simply to drop attempts at regulation, which had been so evidently and frustratingly ineffective in the oil pipeline business, and deregulate the entire 227,000 miles of interstate petroleum pipelines. See Leonard L. Coburn, "The Case for Petroleum Pipeline Deregulation," *Energy Law Journal* 3, no. 1 (1982): 225–72. Chapter 4 described the nontraditional methods, including market-based regulated rates, to which the FERC turned.

49. TAPS is owned by five integrated oil companies: BP, ConocoPhillips, ExxonMobil, Koch, and Unocal.

50. See FERC Opinion No. 502 (123 FERC ¶ 61,287), June 20, 2008. Richard Rapp and I were witnesses for the state of Alaska in that proceeding in hearings held in late 2006. The case dealt with rates that had been achieved by settlement in 1985, in a case in which our late colleagues Herman Roseman and Bruce Netschert testified for the US Department of Justice. *Trans Alaska Pipeline System*, 33 FERC ¶ 61,064, *reh'g denied*, 33 FERC ¶ 61,392 (1985). In short, the issue was whether those "settlement" rates of approximately four dollars per barrel should be continued or whether new rates of approximately two dollars, based on a new cost-of-service study, should replace them. The FERC found for the independent shippers and Alaska, and lowered the rates—an outcome that survived a lengthy appeals process on the part of the pipeline, the final decision coming in 2010 by the US Court of Appeals for the District of Columbia Circuit (no. 08–1270, decided Dec. 3, 2010).

51. Those royalties are computed on the basis of the market price of oil minus the shipping cost from the North Slope to the marine terminal at Valdez.

Chapter 7

1. Organized trading exists not only at the Henry Hub but also at myriad major points of pipeline interconnection throughout the continent. Data from NYMEX and the International Energy Agency show that as of the end of 2010, the volume of forward gas trades at NYMEX was more than 850 times the level of such trades in the countries of continental OECD Europe on the European Energy Exchange (EEX) on roughly equivalent volumes.

2. Troxel, "Long-Distance Natural Gas Pipe Lines," 347.

3. Hooley, *Financing the Natural Gas Industry*, 31.

4. J. C. Youngberg, *Natural Gas, America's Fastest Growing Industry* (San Francisco: Schwabacher-Frey Company, 1930), 21. Youngberg was listed as a partner in Goldman, Jacobs & Co., of San Francisco. This investment report has a chapter entitled "Companies That Are Well Balanced, Integrated Units."

5. Charles Phillips provides an excellent discussion of the abuses of the holding company structure among utilities in the United States by the mid-1930s. See Charles F. Phillips Jr., *The Regulation of Public Utilities* (Arlington, VA: Public Utilities Reports, 1993), 625–35.

6. See Richard D. Cudahy and William D. Henderson, "From Insull to Enron: Corporate (Re)regulation after the Rise and Fall of Two Energy Icons," *Energy Law Journal* 26, no. 1 (2005): 35–110.

7. Troxel, *Economics of Public Utilities*, 165.

8. Sanders, *Regulation of Natural Gas*, 28, 33–34; and Castaneda, *Invisible Fuel*, 107.

9. 49 Stat. 803 (1935). Title I of the act declared utility holding companies to be "affected with the national interest," which required federal regulation. Titles II and III dealt with the regulatory vacuum associated with interstate electricity and gas sales, respectively. Title II brought electricity under the jurisdiction of the Federal Power Commission (FPC), which theretofore had licensed hydroelectric power plants on federal lands and waterways. See Sanders, *Regulation of Natural Gas*, 35. On June 11, 1935, the Senate passed the Public Utility Holding Company Act by a vote of 56–32, despite fierce campaigning against the bill by utility holding company owners and managers. The House of Representatives initially refused to support the Senate bill, but, with amendments, it passed the House and was signed into law by President Franklin Roosevelt. The act was not repealed until the enactment of the Energy Policy Act (EPACT) of 2005 (Section 1263), only after the gas pipeline transport system in the United States had become fully competitive.

10. This was known as the "death sentence clause." But whether the clause actually had teeth in breaking up holding companies was questionable until the Supreme Court used the clause in a ruling: The North U.S. Company v. Securities and Exchange Commission, 90 L. ed. 737, 66 S. Ct. 785 (1946). In this case,

the Supreme Court found in favor of the SEC against North U.S., a holding company. See William H. Anderson, "Public Utility Holding Companies: The Death
Sentence and the Future," *Journal of Land and Public Utility Economics* 23,
no. 3 (1947): 244–54.

11. Phillips, *Regulation of Public Utilities*, 634.

12. Troxel, *Economics of Public Utilities*, 172.

13. Breaking up the holding company system and forcing financial reorganizations on the remaining holding companies, the SEC is an
uncommonly powerful regulator. U.S. governments, functioning
in an environment where the tradition of private enterprise still is
strong, rarely liquidate or reorganize an established industry. The
[alcohol] prohibition amendment destroyed investment values in
the liquor industry, and antitrust legislation forced financial and
marketing reorganizations in a few industries. But no other legislation ever equaled the Holding Company Act in bringing about
broad managerial and financial changes. No law ever forced the di-
vestment of about $9,000,000,000 worth of property. Because Con-
gress saw uncommonly bad practices by holding-company financiers, it was provoked to pass extraordinary legislation. Congress
dealt bluntly with irresponsible, notably acquisitive conduct, prefer-
ring to rebuild the corporate structures rather than to temper the
practices. (ibid., 187–88)

14. See Hooley, *Financing the Natural Gas Industry*, 34–35. There were specific, relatively uncommon, exemptions granted to the PUHCA, such as that allowing National Fuel Gas Corporation to remain in control of its highly interconnected and overlapping gas pipeline and gas distribution operations in New
York and Pennsylvania.

15. The state regulation of utilities began in 1906 with the passage of regulatory legislation in Wisconsin and New York. The Wisconsin statute was written
by John R. Commons at the behest of Wisconsin governor (later senator) Robert
LaFollette Sr. The New York statute was developed independently by Charles
Evans Hughes, later a US Supreme Court justice. Both laws were the immediate
result of the 1905–6 National Civic Federation study.

16. Between 1906 and 1935, the mantle of corporate reform had passed from
the Republicans, led by Theodore Roosevelt, to the Democrats, led by his distant (fifth) cousin Franklin Roosevelt.

17. Title II brought interstate electricity regulation under the jurisdiction of
the Federal Power Commission, which had been established in 1920 to license
hydroelectric power plants on federal lands and navigable streams. See Sanders,
Regulation of Natural Gas, 35.

18. As political historian Elizabeth Sanders, of Cornell University, said, "Independent producers, particularly on the Panhandle and Hugoton fields, stood

to benefit most from the bill's common carrier provision. These beneficiaries, however, were too geographically concentrated to exert political pressure sufficient to overcome the bill's perceived liabilities (even though Rayburn was an influential friend)" (ibid., 37).

19. The valuation standard used by the ICC for oil pipelines was to cause problems in that industry up until the jurisdiction for oil pipelines was passed to the FERC by Congress in 1978. At that time, the FERC found the ICC's "valuation" rate base regulation to be unintelligible, subjective, and unworkable, prompting it to throw out the ICC's approach in favor of a strictly cost-based "trended original cost." However, the FERC "grandfathered" a substantial proportion of existing oil pipeline rates for which the retrospective construction of new cost-based accounting records was thought to be simply impractical after the passage of so many years. See FERC Order No. 154-B (31 FERC, ¶ 61,377).

20. The bill also specified "proper and adequate" depreciation rates to be fixed for different classes of property.

21. Troxel, "II. Regulation of Interstate Movements of Natural Gas," 27–28.

22. Sanders, *Regulation of Natural Gas*, 41.

23. Troxel, "II. Regulation of Interstate Movements of Natural Gas," 29, 30.

24. Sanders, *Regulation of Natural Gas*, 40. There is irony here, for when it came to protecting interstate pipelines from competition in practice, the licensing provision did not work as incumbent pipelines (or apparently Dougherty) would have wished. In a number of cases, particularly in the 1980s and 1990s, both federal and state regulators licensed pipelines that provided head-to-head competition with incumbent pipelines in expanding gas markets, giving greater weight to the potential for pipeline rivalry than to arguments from incumbent gas pipeline companies that they could serve those areas more cheaply. In 1987, the Wisconsin Public Service Commission licensed an entrant over the objections of the incumbent, saying, "This ability to choose provides Wisconsin consumers with options and flexibility . . . both next year and fifty years from now." Wisconsin PSC, Docket No. 6650-CG-104 (Dec. 10, 1987), 19. The commission in New York encouraged the same pipeline entry in a case in 1991 (in which Alfred Kahn and I were witnesses for the entrant), saying, "New Yorkers clearly will benefit from the additional gas transportation option provided by Empire [Pipeline]." New York PSC, Opinion No. 91-3 (Mar. 1, 1991), 33. The Federal Energy Regulatory Commission did the same in 1989, approving the desire of a gas distributor (Citizens Gas) to connect to a new interstate pipeline supplier, saying that a "new supply would give Citizens Gas a second pipeline supplier and thus the ability to diversify." 46 FERC ¶ 61,010 (1989), 61,046.

25. Sanders, *Regulation of Natural Gas*, 41–42.

26. "It is hereby declared that the business of transporting and selling natural gas for ultimate distribution to the public is affected with a public interest, and that Federal regulation in matters relating to the transportation of natural gas

and the sale thereof in interstate and foreign commerce is necessary in the public interest." Natural Gas Act of 1938, 52 Stat. (1938), p. 821.

27. Ibid.

28. My use of the term *private carriage* connotes that a pipeline does not have to serve all comers and is not obligated to plan additional capacity to meet anticipated increased demand. In other words, it implies obligations for a pipeline quite unlike those accompanying *common carriage*. Private carriage, which includes the shipping of a pipeline company's own gas, is a more general term than *contract carriage*, and it better suits my purpose of describing how Congress innovated in regulating pipelines in 1938.

29. Natural Gas Act of 1938, 52 Stat. (1938), p. 824.

30. No natural-gas company shall undertake the construction or extension of any facilities . . . to a market in which natural gas is already being served by another natural-gas company . . . until there shall first have been obtained from the Commission a certificate that the present or future public convenience and necessity requires or will require such new construction or operation: . . . *Provided, however,* That a natural-gas company already serving a market may enlarge or extend its facilities for the purpose of supplying increased market demands in the territory in which it operates. (Natural Gas Act of 1938, 52 Stat. [1938], p. 825; emphasis in the original)

31. "All rates and charges made, demanded, or received by any natural-gas company for or in connection with the transportation or sale of natural gas subject to the jurisdiction of the Commission, and all rules and regulations affecting or pertaining to such rates or charges, shall be just and reasonable, and any such rate or charge that is not just and reasonable is hereby declared to be unlawful" (ibid., 822).

32. "Every natural gas company shall make, keep and preserve for such periods, such accounts, records of cost-accounting, . . . and other records as the Commission may by rules and regulations prescribe as necessary or appropriate for the purposes of the administration of this Act. . . . The Commission may prescribe a system of accounts to be kept by such natural-gas companies." Natural Gas Act of 1938, 52 Stat. (1938), p. 825.

33. The ICC's attempts to regulate oil pipeline accounting are widely considered to have been an abject failure, particularly regarding the identification of the rate base. In the first major oil pipeline rate-making proceeding after it had inherited jurisdiction over oil pipelines from the ICC, the FERC wrestled with the problem without success. The US Court of Appeals for the District of Columbia remanded the FERC's attempt to reuse the ICC's rate base method because it "did not offer a reasoned explanation for adhering to an admittedly antiquated and inaccurate [ICC] regulatory formula." See 31 FERC, 61,377 at 61,832 (Opinion 154-B). The FERC's second attempt (called "trended original cost," or

"TOC") succeeded but has not been free of subsequent controversy. The imperfections of early regulatory accounting schemes, including those of the ICC, are described in Troxel, *Economics of Public Utilities*, 121–22; and Phillips, *Regulation of Public Utilities*, 216–21.

34. Indeed, it was in 1912 that the Supreme Court first decided that if a company provides services to the public, then the public in essence owns that company's operational and financial books and records. In a 1912 case involving transporters regulated by the ICC, the US Supreme Court ruled that accounting systems for public utilities were public matters. See Troxel, *Economics of Public Utilities*, 120, citing Interstate Commerce Com. v. Goodrich Transit Co., 224 U.S. 194, 211 (1912). The next year, the court confirmed that the regulation of accounts by a commission was lawful. See Kansas City Southern Ry. Co. v. U.S., 231 U.S. 423, 440–41 (1913), also cited in Troxel, *Economics of Public Utilities*, 120. These were the legal precursors to the Uniform System of Accounts that accompanied the 1938 Natural Gas Act and fully institutionalized the public character of pipeline operating and financial information in the United States.

35. Troxel, "Accounting Control," chap. 6 in *Economics of Public Utilities*, 115–37. Troxel was involved in the regulatory cases in the 1930s that dealt with the development of accounting standards, and he wrote about these issues extensively, both in his text and in his many scholarly articles. Other, more recent contributors have relied heavily on his writings in the area of regulatory accounting. For example, Charles F. Phillips Jr., of Washington and Lee University, cites Troxel almost three dozen times in his own authoritative text. See Phillips, *Regulation of Public Utilities*.

36. Sanders, *Regulation of Natural Gas*, 42.

37. Ibid., 84.

38. Law professor Richard J. Pierce said, "Since the market imperfection related only to the transportation function, there is every reason to believe that [a common carriage requirement instead] would have been effective to avoid all of the abuses documented by the FTC report. . . . With interstate gas pipelines required to provide equal access to their facilities to all third parties, thousands of producers would be free to sell to hundreds of gas distributors and millions of consumers in a perfectly competitive gas sales market." Pierce, "Reconstituting the Natural Gas Industry," 6–7.

39. Smyth v. Ames, 169 U.S. 466, 546–47 (1898). Commons, writing in 1934, said that in *Smyth*, the Supreme Court gave a "perplexing definition of Reasonable Value." Perplexing or not, however, "when once the Court, by this process of due evaluating, has finally decided a dispute, then that decision, under the institutional set-up of America, is the final word, for the time being, on Reasonable Value." Commons, *Institutional Economics*, 683.

40. Troxel, *Economics of Public Utilities*, 290–91. Eli Clemens, of the University of Maryland, had similar opinions: "The rule of *Smyth v. Ames* is ambigu-

ous and without substance. Insofar as the Court gave it content, fair value came to mean present value, and, in effect, the cost of reproduction. But the Court refused to commit itself to any specific formula or theory." Eli W. Clemens, *Economics and Public Utilities* (New York: Appleton-Century-Crofts, 1950), 147.

41. "It is not too much to say that in terms of cost, delay, uncertainty, and the arousing of animosity and contention, the performance of the . . . method falls little short of a public scandal; by far the greater part of the grotesque and costly ponderosity which characterizes modern rate regulation is to be attributed directly and solely to [that] approach." Leverett S. Lyon and Victor Abramson, *Government and Economic Life: Development and Current Issues of American Public Policy*, vol. 2 (Washington, DC: Brookings, 1940), 691. In a prefatory note to volume 2 of this work, Lyon and Abramson acknowledge that Ben Lewis prepared the chapter from which these quotes were taken. In a much later case, Alfred Kahn referred to an analogous "combat-by-engineering-and-econometric-models." Alfred E. Kahn, *Letting Go: Deregulating the Process of Deregulation*, MSU Public Utility Papers (East Lansing: Michigan State University, 1998), 93.

42. Federal Power Commission et al. v. Hope Natural Gas Co., 320 U.S. 591 (1944), p. 603.

43. Ibid., 602.

44. Judged by its legal history, the reasonable valuation of public utility property is a tough old bone on which many have chewed without getting good and satisfying results. . . . The meaning of reasonableness, which is always something less than perfectly clear and conclusive in a democratic society, is more confused than crystallized by so many gnawings on the valuation bone. . . . The Supreme Court has, I think, the elements . . . of reasonable regulation in the Hope decision; at least it centers attention on the primary question of reasonable earnings rather than reasonable property values, and it is in a good position to reorient commission behavior in future decisions. (Troxel, *Economics of Public Utilities*, 283–84)

45. "John R. Commons gained much of his fame by his great book, *The Legal Foundations of Capitalism*, in which he traced the development of the theory of administered value in the courts. Today a new chapter must be added, as new accounting methods and new law change the entire nature of public utility values and the means by which they are determined. . . . The substitution of adequate and accurate book records for the tedious and expensive appraisal process cannot help but be a gain for regulation." Clemens, *Economics and Public Utilities*, 187–88.

46. James C. Bonbright, "Utility Rate Control Reconsidered in the Light of the *Hope Natural Gas* Case," *American Economic Review* 38, no. 2 (1948): 465.

47. "Had the [Supreme] Court deliberately set out to defeat the whole pur-

pose of regulation and to make public ownership inevitable, it would hardly have pursued this objective more effectively than by its rulings and dicta on valuation." James C. Bonbright, *The Valuation of Property* (New York: McGraw-Hill, 1937), 2:1154.

48. "In any event, the question why the Supreme Court waited so long before conceding, in language and not just in somewhat hazy rulings on evidence, that 'fair value' for rate-making purposes must be given a special meaning in order to avoid the circular-reasoning fallacy, has become of historical interest only. It now seems generally agreed, at least by all experts, that a 'fair-value' measure of the rate base is not the same thing as a 'fair-value' standard in taxation, in the law of damages, or in most other legal appraisals." James C. Bonbright, *Principles of Public Utility Rates* (New York: Columbia University Press, 1961), 165–66.

49. Canada and the United States share a remarkably similar regulatory mandate, and their "fair and reasonable" standards for utilities returns are almost identical. Indeed, Canada's *Northwestern Utilities v. City of Edmonton* anticipated the landmark US *Hope* case by fifteen years. See Northwest Utilities v. City of Edmonton, S.C.R. 186 (NUL 1929).

50. The *Phillips Petroleum* case began when the Wisconsin Public Service Commission and the city of Detroit petitioned the FPC to assert jurisdiction over sales by Phillips Petroleum Co. to the Michigan-Wisconsin Pipeline Company, the principal supplier of gas to the region at the time. The FPC declined to assert jurisdiction, holding that the production by Phillips was "so closely connected" with the production and gathering process that federal rate regulation would encroach on state jurisdiction over gas production. The state of Wisconsin, along with the cities of Milwaukee and Detroit, appealed the decision to the Court of Appeals for the District of Columbia, which reversed the FPC's decision. The appeals court found that the sales by Phillips took place after the production and gathering process and did not interfere with the producing states' regulation of those activities. The Supreme Court upheld the appeals court decision. See Sanders, *Regulation of Natural Gas*, 95.

51. Pierce, "Reconstituting the Natural Gas Industry," 8.

52. Breyer and MacAvoy, *Energy Regulation by the Federal Power Commission*, 57.

53. An important element of the 1942 amendment secured the rights of existing customers to pipeline capacity. In the amendment, Congress specified that once a gas pipeline established service to a customer, it could not expand its system or sell to another customer in a way that would jeopardize its ability to serve its existing customers. In a case involving Panhandle Eastern Pipe Line Company, the FPC did not allow the company to serve the Ford Motor Company, because it did not have "sufficient capacity to sell a large quantity of gas to a new customer without impairing . . . service to existing customers." Troxel, *Economics of Public Utilities*, 96.

54. Joel B. Dirlam, "Natural Gas: Cost, Conservation, and Pricing," *American Economic Review: Papers and Proceedings* 48, no. 2 (May 1958): 492.

55. See what became known as the "omnibus" hearings on gas price regulations before the Federal Power Commission: *In the Matter of Champlin Oil & Refining Co. et al.*, Docket No. G-9277 (1959).

56. Ibid. (testimony of Dr. Alfred E. Kahn), 70–71. In his comments to this writer, Kahn maintained that securing reserves sufficient to enable pipeline promoters to get FPC certification was for them a "license to coin money," which conferred great market power on the producers that were in a position to lease large blocks of reserves.

57. "The bulk of capacity is tied up by long-term contract; only the thin veneer of currently emerging reserves is available to the market in which new prices are determined. On that limited supply converge the ever increasing anticipated requirements of the next twenty years or so, in the demand of pipelines seeking the long-term commitments that are for them the ticket to certification." Alfred E. Kahn, "Economic Issues in Regulating the Field Price of Natural Gas," *American Economic Review: Papers and Proceedings* 50, no. 2 (May 1960): 508–9.

58. The two-tier pricing of natural gas would seem to be contrary to the principles of marginal cost pricing for which Kahn became justly famous. Kahn stated the following regarding its introduction: "The justification [for two-tier pricing], proffered by this writer and accepted by the FERC, was that it was both undesirable and unnecessary to extend that higher price to old gas—undesirable because to do so would confer windfalls on the owners of reserves discovered and developed at lower costs in the past (a noneconomic argument), and unnecessary because the investments in the old gas had already been made (an economic consideration)." Kahn, *Economics of Regulation*, 1:43n55.

59. MacAvoy, *Natural Gas Market*, 57.

60. See Dirlam, "Natural Gas," 491–501. Both Joel Dirlam and Alfred Kahn consulted through the firm of Boni, Watkins, Jason and Co., the precursor to NERA, for the gas distributors battling wellhead price decontrol.

61. Ibid., 494. Dirlam also said that "the 'lure of the big strike' will continue, as it has in the past, to weigh more heavily than changes in price, particularly with independent wildcatters" (ibid.).

62. As it turns out, J. Paul Getty was never a connoisseur of twentieth-century art, and his namesake museum in California has no Jackson Pollocks. It is rare enough to find any mistakes—even in a merely hypothetical example—in my late colleague Joel Dirlam's writings.

63. Kahn, "Economic Issues in Regulating the Field Price of Natural Gas," 506–17.

64. The US Department of Energy calculated that the shortage attributed to price regulation cost consumers between $2.5 and $5.0 billion per year in the form of increased energy costs and lost industrial production. MacAvoy, using

a supply/demand model, estimated that consumers as a group lost more than $20 billion over the period 1968–77. See Pierce, "Reconstituting the Natural Gas Industry," 10; and MacAvoy, *Natural Gas Market*, 15.

65. Spluttering with righteous ire, Simon said, "It is with inexpressible indignation that I report . . . our next energy crisis, a frightening shortage of natural gas. . . . Natural gas was concentrated, of course, in those states where the free market price existed and ran short in those states where federal price controls had reduced supply. . . . Had anything been learned at all?" William E. Simon, *A Time for Truth* (New York: McGraw Hill, 1978), 81–82. Paul MacAvoy used this quote in his dedication to Simon in his 2000 book *The Natural Gas Market*.

66. As cited in Sanders, *Regulation of Natural Gas*, 148–49.

67. An extensive analysis of the origin and politics of the 1978 Natural Gas Policy Act (NGPA) appears in Sanders, *Regulation of Natural Gas*, chap. 7, 165–92.

68. Phillips provides an excellent summary of the complicated "old" and "new" pricing scheme in the NGPA. There was a wide and complicated range of prices. Gas wells drilled ("spudded") before 1973 were to remain regulated at $0.295 per million Btus. "Stripper" wells were to remain regulated at $2.09 per million Btus. Prudhoe Bay (Alaska North Slope) gas was to remain regulated at $1.45 per million Btus. Various types of fields were to be deregulated on various dates, including new onshore and offshore wells (initially regulated at $1.75 per million Btus and deregulated in 1985). It takes Phillips three pages of concise tables to capture all of the "old" and "new" complexity inherent in the NGPA of 1978. See Phillips, *Regulation of Public Utilities*, 500–502.

69. By 1985, industrial users comprised approximately 44 percent of gas consumed in the United States, and about one-third of that industrial demand was for electricity generation. See *Gas Facts 1985: A Statistical Record of the Gas Utility Industry* (Arlington, VA: American Gas Association, 1986), 67.

70. See "Pipeline Take or Pay Costs Continue to Mount," *Oil & Gas Journal*, Aug. 10, 1987, 20. The take-or-pay obligations varied widely among pipeline companies. Many pipelines had limited take-or-pay exposure, due in large part to a sufficiently competitive weighted average cost of gas and/or because of more flexible provisions in their sales gas portfolios.

71. Under Order No. 436, pipelines offering transportation services were required to provide such services on a nondiscriminatory basis in light of existing contracts for firm service. In other words, the same transportation service is to be provided to all shippers willing to pay the applicable tariff rate on file with the FERC. 50 Fed. Reg. 42,408, 42,409 (Oct. 18, 1985).

72. The FERC was explicit in spelling out the voluntary nature of the choice. Because the Natural Gas Act did not treat gas pipelines as carriers for third-party supplies (except for shipments to industrial gas end users), the FERC presumed that it lacked the authority to impose this type of service itself without

a legal challenge that such a move violated the US Constitution's prohibition against harming investors' property without due process and compensation.

73. The FERC recognized that if pipeline companies elected open-access status, they could experience increased take-or-pay liabilities. If an open-access pipeline held contracts with high take-or-pay terms, it would be at a disadvantage vis-à-vis other sellers for gas when competing for gas sales. Order No. 436 offered a means for pipelines to spread part of their take-or-pay liabilities through fixed surcharges to their customers, if those pipelines embraced open access. A succeeding Order No. 500 was required to address legal obstacles to the implementation of this policy, offering pipeline companies a mechanism to recover roughly half of their uneconomic gas costs.

74. To the extent that contract shippers continued to pay the FERC-authorized maximum pipeline rate and term under their transport contracts (which are generally well below the value of capacity in the market because of historical cost accounting), they could renew their contracts without limit.

75. 59 FERC 61,030, 18 CFR Part 284 (Order No. 636), Apr. 8, 1992, p. 28.

76. Ibid., 31–32.

77. After Orders No. 436 (1985) and No. 636 (1992), a number of cases before the FERC dealt with whether pipeline companies would be certified to assess an "inventory charge" to their users. In objecting to such a charge, many distributors commented on the way in which pipeline companies' ability to offer "delivered gas" at the city gate displayed the inequality between pipeline-affiliate gas sales and third-party sales. I offered evidence on behalf of a group of distributors in one such case concerning Southern Natural Gas Company. There, the pipeline company was able to pass through to its firm customers weighted average gas costs that were $0.17 to $0.77 cents higher than Louisiana Gulf Coast spot prices delivered to the pipeline. Yet few of those customers switched to the less expensive third-party suppliers. (See Docket No. CP89–1721-000.) The premium represented the ability of pipelines like Southern to extract price concessions owing to the implied flexibility and reliability that pipeline-affiliated gas offered, vis-à-vis third-party gas.

78. "No-notice" service had been an issue that pitted the interests of midwestern gas distributors against those in New England. The latter group of distributors, reflecting their historical position at the end of the line of two major pipeline companies, had always needed to give notice provisions to their pipeline suppliers, whereas the midwestern distributors, served by multiple major pipelines, had not. The midwestern distributors claimed that they needed such a service, and the FERC obliged them by requiring that pipeline companies make a separate charge for the service. That provided incentives for various distributors—notably, the Northern Indiana Public Service Company. Northern Indiana's inventive vice president, Bill Hitchcock, led the way by purchasing shares in Louisiana salt-dome storage projects in order to avoid (through

synchronized injections/withdrawals upstream to match uncertain withdrawals downstream) the "no-notice" imbalance charges at a fraction of the cost-of-service price that his pipeline suppliers had offered under the FERC's requirement. Others followed. In a way, the FERC facilitated competition by requiring pipeline companies to "unbundle" their various transport services.

79. The FERC described the pooling points as follows in its 1992 order: "The FERC believes that the meeting of gas purchasers and gas sellers can be facilitated by the creation of production area pooling areas on individual pipelines. Production area pooling areas may facilitate the aggregation of supplies by all merchants. The pooling areas may either be places where title passes from the gas merchant to the shipper or they may be places where aggregation and balancing and penalties are determined ['paper' pooling points]. The FERC will not mandate pooling areas, but will not permit actions that inhibit their development." 59 FERC 61,030, 18 CFR Part 284 (Order 636), p. 108. When I asked him who was responsible for this insight, Branko Terzic, an FERC commissioner at the time, credited the late chairman of the FERC, Martin Allday (a Texas lawyer appointed by George H. W. Bush) with the vision that resulted in the imposition of the "pooling point" concept.

80. Leading up to the 1992 FERC order, the New York Mercantile Exchange (NYMEX) in 1990 established a futures market using the Henry Hub, a point on the US gas pipeline system in Erath, Louisiana. Shippers and marketers began to use the capacity release mechanisms, particularly after the 1992 order, as an alternative to obtaining transportation from the pipeline companies. The use of NYMEX as the trading vehicle for gas supplies grew rapidly after the 1992 order.

81. The FERC said in its own summary (90 FERC 61,109, CFR Parts 154, 161, 250, and 254 [Order No. 637], Feb. 9, 2000), "In this rule, the FERC is revising its current regulatory framework to improve the efficiency of the market and provide captive customers with the opportunity to reduce their cost of holding long-term pipeline capacity while continuing to protect against the exercise of market power."

82. On most pipelines, the compression turbines employed by pipeline companies to maintain pressure and capacity draw upon the gas itself, which is paid for by shippers in kind. That is, the pipeline company delivers some fraction of the gas tendered to it (say, 96 percent), with the rest going to fuel the compressors needed to overcome friction in the line. This clever device means that pipeline companies do not have to separately purchase fuel for their compressors—keeping them even further removed from the gas commodity market.

83. Two of the oldest interstate gas pipeline companies handled the issue quite differently. Tennessee Gas Transmission Company traditionally rolled-in all of its capacity additions, including its pipeline extension into New England. Texas Eastern Transmission Company, on the other hand, segregated its incremental

construction costs by tying its own incremental construction costs to new capacity contracts and by creating a wholly owned affiliate (Algonquin Gas Transmission Company) to extend service into New England. The FPC, in the era before capacity rights were transferred to shippers, was content with both types of pricing for capacity additions.

84. The FERC had promulgated an incremental rule in some cases (Great Lakes Transmission, Docket No. RP91–143-000 et al., Oct. 31, 1991) while acquiescing to a rolled-in rule in others (Battle Creek Gas Co. v. FPC, 281 F.2nd 42 (D.C. Circuit, 1960).

85. Statement of Policy, PL94–4 (May 31, 1995). I was highly critical of this policy statement, predicting that it would prompt multiple small pipeline proposals to beat the FERC's 5 percent threshold and so cumulatively work to destroy competition in the new pipeline market. "FERC Takes the Wrong Path in Pricing Policy," *Natural Gas* (Wiley) 12, no. 2 (Sept. 1995): 7–11.

86. *Policy Statement on Determination of Need*, 1902-AB86, FERC Docket No. PL-3–000.

87. The Commission finds that the disclosure of detailed transactional information is necessary to provide shippers with the price transparency they need to make informed decisions, and the ability to monitor transactions for undue discrimination and preference. Shippers need to know the price paid for capacity over a particular path to enable them to decide, for instance, how much to offer for the specific capacity they seek. . . . The disclosure of all transactional information without the shipper's name will be inadequate for other shippers to determine whether they are similarly situated to the transacting shipper for purposes of revealing undue discrimination or preference. . . . Finally, to be meaningful, for decision making purposes, the transactional information must be reported at the time of the actual transaction." (Order No. 637, 183–85)

88. The FERC said, "Since electronic bulletin boards have become standard industry-wide practice, the Commission has designed a rule that builds upon their use and sees no new burden in this requirement. Electronic bulletin boards in particular will be required to comply with the new capacity releasing requirement." 59 FERC ¶ 61,030 (1992), p. 70.

89. The fully evolved nature of gas pipeline regulations in the United States was amply demonstrated in June 2008 by FERC Order No. 712, in which the agency displayed its satisfaction with the competitiveness of the market in entitlements. It permanently eliminated any cap on the prices at which legal gas transport entitlements trade in the market. It also facilitated the assignment of entitlements to competitive aggregators for the purpose of more efficiently selling transport rights in a competitive market. See 123 FERC ¶ 61,286 (issued June 19, 2008).

90. I presented a fuller analysis of how the transport system responded to these shocks, with supporting data, in "Seeking Competition and Supply Security in Natural Gas: The US Experience and the European Challenge," in *Security of Energy Supply in Europe*, ed. François Lévêque, Jean-Michel Glachant, Julián Barquín, Christian von Hirschhausen, Franziska Holz, and William J. Nuttal (Cheltenham, UK: Edward Elgar, 2010), 25–28.

91. See William Trapmann and James Todaro, "Natural Gas Residential Pricing Developments during the 1996–97 Winter," *Natural Gas Monthly* (US Energy Information Administration), Aug. 1997.

92. US Energy Information Administration, *Natural Gas Weekly Update*, Dec. 18, 2000. El Paso's gas sales affiliate had acquired rights to substantial capacity in El Paso's pipeline under three contracts from March 2000 to June 2001. In 2002, the chief administrative law judge of the FERC, Curtis L. Wagner Jr., found that El Paso "exercised market power by withholding substantial volumes of capacity to its California delivery points, which tightened the supply and broadened the basis differential." Docket No. RP00–241-006 (Chief Judge's Certification of Record and Initial Decision), 23. El Paso agreed to pay upwards of two billion dollars to the state of California and other private litigants to settle civil lawsuits stemming from this period when its gas sales affiliate controlled capacity in its pipelines to California.

93. US Energy Information Administration, *Natural Gas Weekly Update*, Sept. 29, 2005.

94. I served as a witness in many proceedings for these gas distributor pressure groups before the FERC in the 1980s and 1990s.

95. The group included the gas distributors from greater New York City (Brooklyn Union Gas and Consolidated Edison), the Philadelphia Gas Works, and other gas distributors in the mid-Atlantic states.

96. In 1955, [we] received a call from one George Tyler of the Philadelphia Electric Company, asking for a meeting to discuss a matter concerning the oil industry. . . . When we met, Tyler explained to us that he represented a group of combination utilities as well as straight gas companies who were interveners in a large number of proceedings being held by the . . . FPC . . . as a result of the Supreme Court's 1954 *Phillips* decision interpreting the Natural Gas Act. The companies in Tyler's group were faced with the following situation. On one side in these proceedings were the producers of natural gas, principally the country's largest oil companies. On the other side was the staff of the FPC. . . . Tyler's concern was that the producers were presenting cases that called for unfairly high prices and that the staff of the FPC was ill equipped, professionally, to grapple with the testimony being presented by the producers' witnesses. What Tyler was looking for was an economist or economists

who were knowledgeable in the oil industry but were not beholden
to that industry, who would be willing to respond to the oil companies' presentations. (Jules Joskow, *NERA: A Somewhat Personal
History* [White Plains, NY: National Economic Research Associates, 1990], 2–3)

Jules Joskow was one of the founders of NERA in 1961 and later president of
that firm of consulting economists.

97. According to Mancur Olson:

The basis for Commons' thinking was the view that the market
mechanisms did not of themselves bring about fair results to the
different groups in the economy, and the conviction that this unfairness was due to disparities in the bargaining power of these different groups. These disparities would not be removed by collective
action promoted by the government unless pressure groups forced
through the necessary reforms. . . . The economist, Commons believed, should not look for economic legislation that would be in
the interest of the whole of society; he should rather attach himself to some pressure group or class and counsel it on the measures
that were in its long-run interest. (Olson, *Logic of Collective Action*, 116–17)

98. Williamson, "New Institutional Economics." 596.

Chapter 8

1. Commons, *Economics of Collective Action*, 21.

2. It is important to remember that open access on the gas pipeline system in
the United States was a *voluntary act*, even if, as one judge noted in the context
of the distress of gas pipelines in the 1980s, the extent to which genuine volition
was involved was overstated: "When a condemned man is given the choice between the noose and the firing squad, we do not ordinarily say that he has 'voluntarily' chosen to be hanged." Associated Gas Distributors v. FERC (824 F.2d
981, June 1987), p. 82.

3. See Barr, "Growing Pains," 49–55.

4. The consent decree limiting dividends to 7 percent of rate base led integrated pipeline companies to seek highly leveraged capital structure through
parent debt guarantees in an effort to evade the cap. Some pipelines, such as
Trans Alaska, thus until recently claimed that those guarantees required the use
of parents' capital structures for rate-making purposes (which, as unregulated oil
companies, were mostly equity and therefore expensive for oil pipeline ratepayers). The FERC finally dispensed with that argument in its 2008 Trans Alaska

rate order, using a traditional, rather equally weighted, regulated debt/equity capital structure. See 123 FERC ¶ 61,287, Opinion No. 502 (June 20, 2008).

5. Included here are the questions of who will fund pipeline expansions. Oil pipelines continue to roll-in expansion costs, as it is hard to create incremental prices for a common carrier pipeline, where the default arrangement is a single pot of costs. Thus, pipeline expansion proposals, which have generally become uncontroversial and noncontentious pricing matters for gas pipelines, will continue to be a problem between shippers (integrated versus nonintegrated) on oil lines.

6. Canadian Government, "Agreement among the Governments of Canada, Alberta, British Columbia and Saskatchewan on Natural Gas Markets and Prices," Oct. 31, 1985. Canada complemented this free-market approach by appointing Roland Priddle, a conservative Scottish economist, to chair the National Energy Board. Prior to chairing the NEB, Priddle had worked for the international Royal Dutch Shell Company. In a notable speech made upon his appointment, Priddle stated that "history had proved that the life expectancy of an energy policy stands in inverse relationship to its length and complexity." Peter Mckenzie-Brown, Gordon Jaremko, and David Finch, *The Great Oil Age* (Calgary: Detselig Enterprises, 1993), 141.

7. National Energy Board, *Natural Gas Market Assessment: Ten Years after Deregulation* (Calgary, Alberta: National Energy Board, Nov. 1996), v–viii.

8. Ibid., 22.

9. Ibid. The NEB also removed price caps on the rates charged for capacity released in the secondary market in 1995.

10. See, for example, Alberta's *Generic Cost of Capital* decision, in which the Energy and Utilities Board (EUB) stated, "The Board concurs that the above decisions [*Northwestern*, *Hope*, and *Bluefield*] are the most relevant judicial authorities with respect to the establishment of a fair return for regulated utilities." Alberta Energy and Utilities Board, *Generic Cost of Capital* Decision 2004–052 (2005), 13.

11. Roger A. Morin, *New Regulatory Finance* (Vienna, VA: Public Utilities Reports, 2006), 12.

12. National Energy Board, *Natural Gas Market Assessment*, 22.

13. The NEB does not have a practical rule requiring incremental pricing of new capacity. Without such a rule, any pipeline company can petition to roll-in new capacity expansions to spread the cost to existing shippers—a practice no longer available to US pipelines.

14. Since the early 1970s, the great quantity of gas on Alaska's North Slope has attracted interest in a pipeline through Canada to the lower forty eight states. Interest in that pipeline ebbs and flows, peaking with the spike in gas prices in mid-2008, when a preliminary agreement was struck between the state of Alaska

and the TransCanada Pipeline. It remained to be seen, however, whether such interest would continue after gas prices halved later in the year. If such a pipeline were ever to be built, the US gas distribution companies and power-generating plants that would constitute the major firm shippers, and that would be the chief engine for capital repayment for the line, would almost certainly want to own their transport entitlements as they do for the US gas pipelines that serve them now.

15. The original Gas Act of 1986 did specify (Part I, Section 4 [2][d]) that the director general of Ofgas should "enable persons to compete effectively in the supply of gas through pipes at rates which, in relation to any premises, exceed 25,000 therms a year." However, without explicit technical/informational mechanisms or cooperation from British Gas in providing such open access, little success was achieved in this regard.

16. See "The Statement of the Gas Transmission Transportation Charging Methodology" ("Methodology") and "The Statement of Gas Transmission Transport Charges" ("Charges"), effective from Apr. 1, 2007, National Grid.

17. The first version of this model was described in detail to me in 1995 by the staff at British Gas responsible for the model. At that time, it took approximately one man-month to run the model, computing the LRMC for a set of hypothetical capacity additions needed for the individual capacity expansion at each individual entry and exit point, holding all others constant.

18. It has become commonplace for Ofgem, the regulator, to respond to criticisms of the inefficiency of its practices by calling for another auction of some type of service of Ofgem's design. If those auctions attract little interest, then Ofgem may compel participation from shippers or others through its control over shipper licenses "to ensure that all parties participate appropriately." See Ofgem, letter from managing director, Networks, re "DN [distribution network] Interruptions Reform and Shipper Participation," Sept. 12, 2008.

19. The accounting and financial rules for tariff making for privatized companies in the United Kingdom derived from a 1986 document dealing with valuations for tariff making for publicly owned utilities entitled "Accounting for Economic Costs and Changing Prices: A Report to HM Treasury by an Advisory Group." Led by I. C. R. Byatt, then deputy chief economic adviser, HM Treasury, the "Byatt report" is well known in UK regulatory circles. The Byatt report was also an important reference document in the later privatizations in New Zealand and Australia—particularly in the formation of the "optimised deprival value" valuation metric by the New Zealand Ministry of Commerce in 1993 and 1994, which was later effectively adopted, with different terminology, in Australia. See *Rationale for Financial Performance Measures in the Electricity Information Disclosure Regime*, a report to the Energy Policy Group by Ernst & Young, Aug. 1994.

20. With regard to the valuation of property, the Byatt report concluded: "The value of assets to a business means what potential competitors would find it worth paying for them, even if the competition is hypothetical. This will be the net replacement cost of a Modern Equivalent Asset if the asset would be worth replacing, or the recoverable amount if it would not" (1:6). This conclusion reflects a perspective exactly opposite of that expressed by Bonbright in his comprehensive *Valuation of Property* treatise and reflected by the US Supreme Court in its *Hope* decision. It is no wonder that US and UK economists often speak past each other on the question of the foundation for regulated prices.

21. Ofgem relies on accounting consultants at each regulatory review to standardize cost definitions, but each set of accountants is free to adopt new rules. Ofgem has demonstrated that it is willing to disallow past capital expenditures, ostensibly because of inefficiency, but in practice because projects in question were more expensive than forecast. See Ofgem, *Gas Distribution Price Control Review: One Year Control Final Proposals* (London: Ofgem, 2006), chap. 3.

22. See East Australian Pipeline Pty Limited v. Australian Competition and Consumer Commission (2007) HCA 44 (Sept. 27, 2007).

23. It would be an omission not to mention the efforts of the late José "Pepe" Estenssoro, the president of YPF (Yacimientos Petroliferos Fiscales) and thereby the chief executive of Gas del Estado. Estenssoro worked forcefully with Givogre's economists in the government and those hired by the World Bank (including myself) to make possible the rapid restructuring of that state-owned business (into two pipeline companies and eight independent distributors). From the perspective of creating the conditions for successful arm's-length transacting with asset-specific gas pipelines, he was the antithesis of Sir Denis Rooke in the United Kingdom. Estenssoro died when his small YPF jet crashed in Quito, Ecuador, in 1995.

24. The Argentine problems have produced considerable research into the institutional weaknesses exposed by such privatizations, such as limitations in regulatory capacity, accountability, and fiscal efficiency—all critical to the foundation for effective regulation of investor-owned pipeline undertakings. See Antonio Estache and Liam Wren-Lewis, "Toward a Theory of Regulation for Developing Countries: Following Jean-Jacques Laffont's Lead," *Journal of Economic Literature* 47, no. 3 (2009): 729–69.

25. Leading up to the privatization, the initial structure for the tariff-related books was designed by NERA and finalized by Stone and Webster, both of which based the accounts on the US Uniform System of Accounts and the pipeline tariffs themselves on procedures generally in practice at the FERC.

26. I had discovered this for myself when working with one of the major European-owned distribution companies in Buenos Aires six years after the initial privatization. The care and maintenance of those accounting records, which

I had assumed at the time of privatization would readily continue in the years to follow, had largely been abandoned through inattention on the part of the regulatory agency.

27. Carmen M. Reinhart and Miguel A. Savastano, "The Realities of Modern Hyperinflation," *Finance and Development* 40 (June 2003): 21.

28. See Ricardo Hausman and Andrés Velasco, "Hard Money's Soft Underbelly: Understanding the Argentine Crisis," *Brookings Trade Forum* (2002); and J. F. Hormbeck and Meaghan K. Marshall, *The Argentine Financial Crisis: A Chronology of Events*, Report for Congress, Congressional Research Service, June 5, 2003.

29. Anouk Honoré, "Argentina: 2004 Gas Crisis," *Oxford Institute for Energy Studies*, Nov. 2004.

30. In the EU, Russia serves less than 30 percent of the market, and the top five producers account for about 85 percent. See Holz, von Hirschhausen, and Kemfert, "Strategic Model of European Gas Supply."

31. For a description of the evolution of the municipal distributors in the Netherlands, see Peebles, *Evolution of the Gas Industry*, 132–34.

32. Gazprom has investments in pipelines throughout Europe, including the major interconnector pipelines (specifically, the UK Interconnector, the Blue Stream pipeline, the Yamal-Europe pipeline through Poland and Belarus, and the Nord Stream pipeline in the Baltic Sea). Gazprom also has joint ventures with pipelines or national gas suppliers in France, Italy, Germany, Hungary, the Netherlands, Poland, and Switzerland, among other countries.

33. Indeed, Chris Patten expressed just such concerns over Russia's ability to divide EU member states over energy policy, citing former chancellor Gerhard Schröeder's acceptance of the chairmanship of the Gazprom shareholders committee shortly after leaving office, the bilateral deals between Russia and big energy companies in some European countries (France, Austria, Italy, and Germany), and Russia's evident desire to attempt to gain a lock on gas imports in southeastern Europe. See Chris Patten, "What Is Europe to Do?" *New York Review of Books* 57, no. 3 (Mar. 11, 2010): 12.

34. Some European gas pipeline companies (particularly in Germany and the Netherlands) are not state owned, as such, but include large municipal shareholdings.

35. The EU does have a Single Market Treaty, and EU law is supposed to override national law. However, the EU has no independent political mandate and so is dependent upon the agreement of member states for all new laws and regulations. Regulation (EC) No. 713/2009 of the European Parliament and of the Council, of July 13, 2009, does establish an agency for the cooperation of member state energy regulators. But that agency is advisory only.

36. See Tooraj Jamasb, Michael Pollitt, and Thomas Peter Triebs, "Productivity and Efficiency of US Gas Transmission Companies: A European Regulatory

Perspective," *Energy Policy* 36, no. 9 (Sept. 2008): 3399. The authors mention the "often stunning differences in transparency between the US and Europe" (see footnote 2). In Europe, pipelines sometimes publish measures of "available capacity," but this capacity is not consistently defined, and pipeline companies retain considerable control over the figures. Names of shippers are kept secret by the pipeline companies, and information about shipping gas across the continent is unavailable to open scrutiny, imposing a heavy logistical and information burden on European shippers that would independently attempt to ship gas supplies across the continent.

37. In its *Competition Report* of January 2007, it said, "In the absence of any single cross-border regulator, national regulators must cooperate with each other in monitoring the management and allocation of interconnection capacity. . . . Moreover, the matter in which Community rules have been implemented varies between Member States, and may in some cases even give rise to regulatory vacuum—especially in cross border situations." European Commission, *DG Competition Report on Energy Sector Inquiry* (Brussels: European Commission, Jan. 10, 2007), ¶ 50, 59.

38. Directive 2009/73/EC of the European Parliament and of the Council, of July 13, 2009, concerning common rules for the internal market in natural gas and repealing Directive 2003/55/EC ("third legislative package"), Article 42.

39. Treating these terms as synonymous merely reflects common practice in the economic literature on the subject—which is vague on the precise meaning and obligations attached to either term as commonly used. Legal scholars might be expected to object vigorously—with good reason—to the vague equivalence with which economists and others seem to treat these terms.

40. Third legislative package, Directive 2009/73/EC, Article 32.

41. Ibid., preliminary ¶ 23. Not all of the European gas pipeline system is bound by TPA rules. A distinction exists between the national systems and the large, international supply pipelines (the "interconnectors") that transport gas from producing countries that do have long-term contracts. See Directive 2009/73/EC, Article 36.

42. "The current level of unbundling of network and supply interests has negative repercussions on market functioning and on incentives to invest in networks. This constitutes a major obstacle to new entry and also threatens security of supply." European Commission, *DG Competition Report*, 7.

43. "While the Commission considers that ownership unbundling remains the preferred option, it does however provide for an *alternative option* for Member States that choose not to go down this path. . . . This option, a derogation from the basic ownership unbundling approach, is known as the 'Independent System Operator.' This option enables vertically integrated companies to retain the ownership of their network assets." Proposal for a Directive of the European Parliament and of the Council, Amending Directive 2003/55/EC, third legisla-

tive package, Explanatory Memorandum, Brussels, 2007, Section 1.2 (emphasis in the original). In the *Oxford English Dictionary*, 2nd ed., *derogation* means the weakening of authority or the partial repeal of the law. The alternative option is indeed a very great weakening of the requirement for ownership unbundling, even in the possibility for misuse of affiliate transactions. It effectively removes completely the possibility that ownership unbundling could lead to pipeline rivalry in the area served by the pipeline in question. The proposal was accepted as Article 9.8 and Chapter IV in the third legislative package.

44. The Explanatory Memorandum for the third legislative package makes a number of references to the electricity sector.

45. Third legislative package, Articles 16 and 30.

46. The Sector Inquiry confirms that gas wholesale operators have contrasting views on the question whether the amount of information available on network capacity is sufficient. Incumbents are usually satisfied, whereas most new entrants find that information is lacking, suggesting that vertically integrated incumbents have privileged access to information. . . . It may be a concern that excessive transparency could facilitate collusion between the major market players, particularly on an oligopolistic market. A balance must certainly be found as to what data is published and how it is published, in order to improve transparency without enabling collusion. (European Commission, *DG Competition Report*, 90)

47. The perception that public information could support collusive agreements constituted one of the major reasons given by the Australian Competition Tribunal for declining to regulate the Eastern Gas Pipeline, which led to the deregulation of others in Australia. See chap. 4.

48. "The Commission has also gathered indications that one TSO [transmission system operator] grants its affiliated supply company substantive rebates for the transportation fees as compared to non-affiliated network users. In doing so, the TSO directly supports the competitive position of the related supply company. . . . If the ownership link is broken the incentives facing the network operator will change. It will seek to optimise its network business as opposed to acting in the overall interest of the vertically integrated group." European Commission, *DG Competition Report*, 58.

49. Daron Acemoglu, of MIT, a prominent researcher on how constitutions and property rights affect economic growth, writes that there are "many reasons to expect societies that have better property rights institutions to also have better contracting institutions." Daron Acemoglu, "Constitutions, Politics, and Economics: A Review Essay on Persson and Tabellini's *The Economic Effects of Constitutions*," *Journal of Economic Literature* 47, no. 4 (Dec. 2005): 1025–48.

50. See Regulation (EC) No. 715/2009 of the European Parliament and of the Council of July 13, 2009, Article 13: "By 3 September 2011, the Member States

shall ensure that, after a transitional period, network charges shall not be calculated on the basis of contract paths."

51. See the foreword by Jonathan Stern, of Oxford University, in Paul Hunt's working paper "Entry-Exit Transmission Pricing with Notional Hubs: Can It Deliver a Pan-European Wholesale Market in Gas?" Oxford Institute for Energy Studies, Working Paper NG 23, Feb. 2008, ii.

52. See the Explanatory Memorandum for the proposed third package, Section 5.6.

53. Those independent system operators that already exist in the isolated markets of the United Kingdom and Victoria are instructive with respect to their cost, their evident disregard for pipeline rivalry, and their overriding desire to preserve and expand their jurisdiction.

54. As a former US ambassador to Turkey (now a fellow at the Brookings Institution) remarked to me at a 2009 Brookings conference on Turkey, Russia, and regional energy issues, to the extent that institutions prevent supply security in Europe, "the solution would appear to lie not in more pipelines to the east, but in Brussels."

55. Olson, *Rise and Decline of Nations*, 44.

Chapter 9

1. "For economists, if not more generally, governance and organization are important if and as these are made susceptible to analysis. As described here, breathing operational content into the concept of governance would entail examining economic organization through the lens of contract (rather than the neoclassical lens of choice), recognizing that this was an interdisciplinary project where economics and organization theory (and, later, aspects of the law) were joined, and introducing hitherto neglected transaction costs into the analysis." Oliver E. Williamson, "Transaction Cost Economics: The Natural Progression," *American Economic Review* 100, no. 3 (June 2010): 673. This was the revised version of Williamson's Nobel Prize lecture given in Stockholm on Dec. 8, 2009.

Bibliography

Acemoglu, D. "Constitutions, Politics, and Economics: A Review Essay on Persson and Tabellini's *The Economic Effects of Constitutions.*" *Journal of Economic Literature* 47, no. 4 (Dec. 2005): 1025–48.

Acemoglu, D., S. Johnson, and J. A. Robinson. "Institutions as a Fundamental Cause of Long-Run Growth." In *Handbook of Economic Growth*. Edited by A. Aghion and S. N. Durlauf. Vol. 1A. Amsterdam: Elsevier, 2005.

Adelman, M. A. *The World Petroleum Market*. Baltimore: Johns Hopkins University Press, 1972.

Anderson, R. E., and R. T. Rapp. *Competition in Oil Pipeline Markets: A Structural Analysis*. White Plains, NY: National Economic Research Associates, 1978.

Anderson, W. H. "Public Utility Holding Companies: The Death Sentence and the Future." *Journal of Land and Public Utility Economics* 23, no. 3 (1947): 244–54.

Armstrong, M., S. Cowan, and J. Vickers. *Regulatory Reform, Economic Analysis and the British Experience*. Cambridge, MA: MIT Press, 1994.

Arrow, K. J. "Reflections on the Essays." In *Arrow and the Foundations of the Theory of Economic Policy*, edited by G. Feiwel, 727–34. New York: New York University Press, 1987.

Bailey, E. E., D. R. Graham, and D. P. Kaplan. *Deregulating the Airlines*. Cambridge, MA: MIT Press, 1985.

Bajari, P., and S. Tadelis. "Incentives versus Transaction Costs: A Theory of Procurement Contracts." *RAND Journal of Economics* 32, no. 3 (Autumn 2001): 387–407.

Baker, G., R. Gibbons, and K. J. Murphy. "Relational Contracts and the Theory of the Firm." *Quarterly Journal of Economics* 117, no. 1 (Feb. 2002): 39–84.

Baron, D. P., and J. A. Ferejohn. "Bargaining in Legislatures." *American Political Science Review* 83, no. 4 (Dec. 1989): 1181–1205.

Barr, C. J. "Growing Pains: FERC's Reponses to Challenges to the Development of Oil Pipeline Infrastructure." *Energy Law Journal* 28, no. 1 (2007): 43–70.

Baumol, W. J., J. C. Panzar, and R. D. Willig. *Contestable Markets and the Theory of Industrial Structure.* New York: Harcourt Brace Jovanovich, 1982.

Beckjord, W. C. *"The Queen City of the West"—During 110 Years! A Century and 10 Years of Service by the Cincinnati Gas & Electric Company, 1841–1951.* New York: Newcomen Society in North America, 1951.

Belkin, P. "CRS Report for Congress: The European Union's Energy Security Challenges." *Statistical Pocket Book, 2007.* Washington, DC: Congressional Research Service, Jan. 30, 2008.

Bertrand, J. "Review of *Théorie mathematique de la richesse sociale* and *Recherches sur les principles mathematique de la theorie des richesses.*" *Journal des Savants* (1883): 499–508.

Bonbright, J. C. *Principles of Public Utility Rates.* New York: Columbia University Press, 1961.

———. "Utility Rate Control Reconsidered in the Light of the *Hope Natural Gas Case.*" *American Economic Review* 38, no. 2 (May 1948): 465–82.

———. *The Valuation of Property.* 2 vols. New York: McGraw-Hill, 1937.

Breyer, S. G., and P. W. MacAvoy. *Energy Regulation by the Federal Power Commission.* Washington, DC: Brookings, 1974.

Castaneda, C. J. *Invisible Fuel: Manufactured and Natural Gas in America, 1800–2000.* New York: Twayne Publishers, 1999.

———. *Regulated Enterprise: Natural Gas Pipelines and Northeastern Markets, 1938–1954.* Columbus: Ohio State University Press, 1993.

Castaneda, C. J., and C. M. Smith. *Gas Pipelines and the Emergence of America's Regulatory State: A History of Panhandle Eastern Corporation, 1928–1993.* Cambridge: Cambridge University Press, 1996.

Cheung, S. N. S. "Ronald Henry Coase (b. 1910)." In *The New Palgrave: A Dictionary of Economics.* 1st ed. London: Macmillan, 1987, 1:455–57.

Christensen, L. R., and W. H. Greene. "Economies of Scale in U.S. Electric Power Generation." *Journal of Political Economy* 84, no. 4 (Aug. 1976): 655–76.

Clemens, E. W. *Economics and Public Utilities.* New York: Appleton-Century-Crofts, 1950.

Coase, R. H. "Coase on Posner on Coase." *Journal of Institutional and Theoretical Economics* 149, no. 1 (1993): 96–98.

———. "The Federal Communications Commission." *Journal of Law and Economics* 2 (1959): 1–40.

———. "The Nature of the Firm." *Economica* 4, no. 16 (1937): 386–405.

———. "The Problem of Social Cost." *Journal of Law and Economics* 3 (1960): 1–44.

Coburn, L. L. "The Case for Petroleum Pipeline Deregulation." *Energy Law Journal* 3, no. 1 (1982): 225–72.

Commons, J. R. *The Economics of Collective Action.* New York: Macmillan, 1950.

———. *Institutional Economics.* New York: Macmillan, 1934.

———. *Legal Foundations of Capitalism.* New York: Macmillan, 1924.

———. *Myself.* New York: Macmillan, 1934.

Cookenboo, L., Jr. *Crude Oil Pipe Lines and Competition in the Oil Industry.* Cambridge, MA: Harvard University Press, 1955.

Crocker, K. J., and S. E. Masten. "Regulation and Administered Contracts Revisited: Lessons from Transaction-Cost Economics for Public Utility Regulation." *Journal of Regulatory Economics* 8 (1996): 5–39.

Crocker, K. J., and K. J. Reynolds. "The Efficiency of Incomplete Contracts: An Empirical Analysis of Air Force Engine Procurement." *RAND Journal of Economics* 24, no. 1 (Spring 1993): 126–46.

Cudahy, R. D., and W. D. Henderson. "From Insull to Enron: Corporate (Re) regulation after the Rise and Fall of Two Energy Icons." *Energy Law Journal* 26, no. 1 (2005): 35–110.

Daggett, S. R. *Principles of Inland Transportation.* New York: Harper and Brothers, 1928.

Dales, J. H. *Pollution, Property and Prices.* Toronto: University of Toronto Press, 1968.

Dalton, J. A., and L. Esposito. "Predatory Price Cutting and Standard Oil: A Re-examination of the Trial Record." *Research in Law and Economics* 22 (2007): 155–205.

Davis, L. E., and D. C. North. *Institutional Change and American Economic Growth.* Cambridge: Cambridge University Press, 1971.

Davison, A., C. Hurst, and R. Mabro. *Natural Gas: Governments and Oil Companies in the Third World.* Oxford Institute of Energy Studies. Oxford: Oxford University Press, 1988.

de Chazeau, M. E., and A. E. Kahn. *Integration and Competition in the Petroleum Industry.* New Haven, CT: Yale University Press, 1959.

DiLorenzo, T. J. "The Myth of Natural Monopoly." *Review of Austrian Economics* 9, no. 2 (1996): 43–58.

Dirlam, J. B. "Natural Gas: Cost, Conservation, and Pricing." *American Economic Review: Papers and Proceedings* 48, no. 2 (May 1958): 491–501.

Dixit, A. "Governance Institutions and Economic Activity." *American Economic Review* 99, no. 1 (Mar. 2009): 5–24.

Dixon, F. H. "The Mann-Elkins Act, Amending the Act to Regulate Commerce." *Quarterly Journal of Economics* 24, no. 4 (Aug. 1910): 593–633.

Ellerman, A. D., P. L. Joskow, and D. Harrison Jr. "Emissions Trading in the US: Experience, Lessons and Considerations for Greenhouse Gases." Pew Center on Global Climate Change, May 2003.

Erize, L. A., and S. M. Porteiro. "Argentina." *Gas Regulation 2007.* London: Global Legal Group, 2007.

Estache, A., and L. Wren-Lewis. "Toward a Theory of Regulation for Developing Countries: Following Jean-Jacques Laffont's Lead." *Journal of Economic Literature* 47, no. 3 (2009): 729–69.

European Commission. *DG Competition Report on Energy Sector Inquiry.* Brussels: European Commission, Jan. 10, 2007.

Furubotn, E. G., and R. Richter. *Institutions and Economic Theory: The Contributions of the New Institutional Economics.* Ann Arbor: University of Michigan Press, 1997.

Ganguin, B. *Fundamentals of Corporate Credit Analysis*. New York: McGraw Hill, 2005.

Glaeser, M. G. *Outlines of Public Utility Economics*. New York: Macmillan, 1927.

Goldberg, V. P. "Pigou on Complex Contracts and Welfare Economics." *Research in Law and Economics* 39 (1981): 39–51.

Grossman, S. J., and O. D. Hart. "The Costs and Benefits of Ownership: A Theory of Vertical and Lateral Integration." *Journal of Political Economy* 94, no. 4 (1986): 39–84.

Hadfield, G. K. "The Many Legal Institutions That Support Contractual Commitments." In *Handbook of New Institutional Economics*, edited by C. Ménard and M. M. Shirley, 175–203. Dordrecht, Netherlands: Springer, 2005.

Hansen, J. *U.S. Oil Pipeline Markets*. Cambridge, MA: MIT Press, 1983.

Hart, O. D. "Corporate Governance: Some Theory and Implications." *Economic Journal* 105, no. 430 (May 1995): 678–89.

Hart, O. D., and J. Moore. "Property Rights and the Nature of the Firm." *Journal of Political Economy* 98, no. 6 (1990): 1119–58.

Hausman, R., and A. Velasco. "Hard Money's Soft Underbelly: Understanding the Argentine Crisis." *Brookings Trade Forum* (2002).

Henderson, J. M., and R. E. Quandt. *Microeconomic Theory: A Mathematical Approach*. 2nd ed. New York: McGraw-Hill, 1971.

Hodgson, G. "John R. Commons and the Foundations of Institutional Economics." *Journal of Economic Issues* 37 (Sept. 2003): 547–76.

Holmes, Oliver Wendell. "The Path of the Law," address, 1897. Reprinted in *Collected Legal Papers*. New York: Harcourt, Brace and Howe, 1920.

Holz, F., C. von Hirschhausen, and D. Kemfert. "A Strategic Model of European Gas Supply." Discussion paper 551, DIW Berlin, Jan. 2006.

Honoré, Anouk. "Argentina: 2004 Gas Crisis." Oxford Institute for Energy Studies, Nov. 2004.

Hooley, R. W. *Financing the Natural Gas Industry*. New York: AMS Press, 1968.

Hormbeck, J. F., and M. K. Marshall. *The Argentine Financial Crisis: A Chronology of Events*. Report for Congress, Congressional Research Service, June 5, 2003.

Hunt, P. "Entry-Exit Transmission Pricing with Notional Hubs: Can It Deliver a Pan-European Wholesale Market in Gas?" Oxford Institute for Energy Studies, Working Paper NG 23, Feb. 2008.

Hunt, S., and G. Shuttleworth. *Competition and Choice in Electricity*. New York: Wiley, 1996.

Jamasb, T., M. Pollitt, and T. Triebs. "Productivity and Efficiency of US Gas Transmission Companies: A European Regulatory Perspective." *Energy Policy* 36, no. 9 (Sept. 2008): 3398–3412.

Johnson, A. M. *The Development of American Petroleum Pipelines: A Study in Private Enterprise and Public Policy, 1862–1906*. Ithaca, NY: Cornell University Press, 1956.

———. *Petroleum Pipelines and Public Policy, 1906–1959.* Cambridge, MA: Harvard University Press, 1967.

Joskow, J. *NERA: A Somewhat Personal History.* White Plains, NY: National Economic Research Associates, 1990.

Joskow, P. L. "Transaction Cost Economics, Antitrust Rules, and Remedies." *Journal of Law, Economics and Organization* 18, no. 1 (2002): 95–116.

———. "Vertical Integration." In *Handbook of New Institutional Economics,* edited by C. Ménard and M. M. Shirley. Dordrecht, Netherlands: Springer, 2005.

Kahn, A. E. "Economic Issues in Regulating the Field Price of Natural Gas." *American Economic Review: Papers and Proceedings* 50, no. 2 (May 1960): 506–17.

———. *The Economics of Regulation: Principles and Institutions.* 2 vols. New York: Wiley, 1971.

———. *Letting Go: Deregulating the Process of Deregulation.* MSU Public Utility Papers. East Lansing: Michigan State University, 1998.

———. "Reflections of an Unwitting 'Political Entrepreneur.'" *Review of Network Economics* 7, no. 4 (Dec. 2008): 616–29.

———. "Telecommunications: The Transition from Regulation to Antitrust." *Journal of Telecommunications and High Technology Law* 5, no. 1 (2007): 159–88.

Keeler, T. E. *Railroads, Freight and Public Policy.* Washington, DC: Brookings Institution, 1983.

Kennedy, J. L. *Oil and Gas Pipeline Fundamentals.* Tulsa, OK: PennWell Books, 1993.

Klein, B. J., R. Crawford, and A. Alchian. "Vertical Integration, Appropriable Rents, and the Competitive Contracting Process." *Journal of Law and Economics* 21, no. 2 (1978): 297–326.

Knight, F. H. *Risk, Uncertainty and Profit.* 1921. Reprint, Chicago: University of Chicago Press, 1971.

Kwerel, E. R., and G. L. Rosston. "An Insider's View of FCC Spectrum Auctions." *Journal of Regulatory Economics* 7, no. 3 (May 2000): 253–89.

Landis, J. M. *Report on Regulatory Agencies to the President Elect.* US Senate Committee on the Judiciary, 86th Cong., 2nd Sess. Washington, DC: Committee Print, 1960.

La Porta, R., F. Lopez-de-Silanes, A. Shleifler, and R. W. Visny. "Law and Finance." *Journal of Political Economy* 106, no. 6 (Dec. 1998): 1113–55.

Leeston, A. M., J. A. Crichton, and J. C. Jacobs. *The Dynamic Natural Gas Industry.* Norman: University of Oklahoma Press, 1963.

Lohman, S. "Rational Choice and Political Science." *The New Palgrave Dictionary of Economics,* edited by S. N. Durlauf and L. E. Blume. 2nd ed. New York: Palgrave Macmillan, 2008.

Lyon, L. S., and V. Abramson. *Government and Economic Life: Development and Current Issues of American Public Policy.* Vol. 2. Washington, DC: Brookings, 1940.

Mabro, R., and I. Wybrew-Bond, eds. *Gas to Europe: The Strategies of Four Major Suppliers*. Oxford: Oxford University Press, 1999.

MacAvoy, P. W. *The Economic Effects of Regulation*. Cambridge, MA: MIT Press, 1965.

———. *The Natural Gas Market: Sixty Years of Regulation and Deregulation*. New Haven, CT: Yale University Press, 2000.

———. *Price Formation in Natural Gas Fields*. New Haven, CT: Yale University Press, 1962.

MacAvoy, P. W., and S. Breyer. *Energy Regulation by the Federal Power Commission*. Washington, DC: Brookings Institution, 1974.

Makholm, J. D. "FERC Takes the Wrong Path in Pricing Policy." *Natural Gas* (Wiley) 12, no. 2 (Sept. 1995): 7–11.

———. "Gas Pipeline Capacity: Who Owns It, Who Profits, Who Pays?" *Public Utilities Fortnightly* 132, no. 18 (Oct. 1, 1993): 17–20.

———. "Seeking Competition and Supply Security in Natural Gas: The US Experience and the European Challenge." In *Security of Energy Supply in Europe*, edited by F. Lévêque, J.-M. Glachant, J. Barquín, C. von Hirschhausen, F. Holz, and W. J. Nuttal, 21–55. Cheltenham, UK: Edward Elgar, 2010.

Masten, S. E., and S. Saussier. "Econometrics of Contracts: An Assessment of Developments in the Empirical Literature of Contracting." In *Economics of Contracts: Theories and Applications*, edited by E. Brousseau and J.-M. Glachant, 273–92. Cambridge: Cambridge University Press, 2002.

McAllister, E. W., ed. *Pipeline Rules of Thumb Handbook*. 4th ed. Houston: Gulf Publishing Company, 1998.

McCraw, T. K. *Prophets of Regulation*. Cambridge, MA: Harvard University Press, 1984.

McGee, J. S. "Predatory Price Cutting: The Standard Oil (N.J.) Case." *Journal of Law and Economics* 289 (1958): 137–69.

———. "Predatory Price Cutting Revisited." *Journal of Law and Economics* 289 (1980): 289–330.

Mckenzie-Brown, P., G. Jaremko, and D. Finch. *The Great Oil Age*. Calgary: Detselig Enterprises, 1993.

Ménard, C. "The Economics of Hybrid Organizations." *Journal of Institutional and Theoretical Economics* 160, no. 3 (2004): 345–76.

Ménard, C., and M. M. Shirley, eds. *Handbook of New Institutional Economics*. Dordrecht, Netherlands: Springer, 2005.

Merryman, J. H. "Ownership and Estate (Variations on a Theme by Lawson)." *Tulane Law Review* 48 (June 1974): 916–45.

Morin, R. A. *New Regulatory Finance*. Vienna, VA: Public Utilities Reports, 2006.

Mulherin, J. H. "Complexity in Long-Term Contracts: An Analysis of Natural Gas Contractual Provisions." *Journal of Law, Economics, and Organization* 2, no. 1 (Spring 1986): 105–17.

Munro, W. B. "Review: The Civic Federation Report on Public Ownership." *Quarterly Journal of Economics* 23, no. 1 (1908): 161–74.

Nash, J. F., Jr. "The Bargaining Problem." *Econometrica* 18 (1950): 155–62.

National Civic Federation. *Municipal and Private Operation of Public Utilities.* 3 vols. New York: National Civic Federation, 1907.

Nelson, J. R. "The Role of Competition in Regulated Industries." *Antitrust Bulletin* 11 (1966): 1–36.

Nerlove, M. "Returns to Scale in Electricity Supply." In *Measurement in Economics*, by C. F. Christ, M. Friedman, L. A. Goodman, Z. Griliches, A. C. Harberger, N. Liviatan, J. Mincer, et al. Stanford, CA: Stanford University Press, 1963.

Newbery, D. M. *Privatization, Restructuring and Regulation of Network Utilities.* The Walras-Pareto Lectures, 1995. Cambridge, MA: MIT Press, 2000.

North, D. C. "Economic Performance through Time." *American Economic Review* 84, no. 3 (June 1994): 359–68.

———. "Sources of Productivity Change in Ocean Shipping." *Journal of Political Economy* 76 (1968): 953–70.

Nuechterlein, J. E., and P. J. Weiser. *Digital Crossroads: American Telecommunications Policy in the Internet Age.* Cambridge, MA: MIT Press, 2007.

Olson, M. "Collective Action." *The New Palgrave Dictionary of Economics.* Edited by S. G. Durlauf and L. E. Blume. 2nd ed. New York: Palgrave Macmillan, 2008.

———. *The Logic of Collective Action: Public Goods and the Theory of Groups.* Cambridge, MA: Harvard University Press, 1965.

———. *The Rise and Decline of Nations: Economic Growth, Stagflation, and Social Rigidities.* New Haven, CT: Yale University Press, 1982.

Patten, C. "What Is Europe to Do?" *New York Review of Books* 57, no. 3 (Mar. 11, 2010): 11–12.

Peebles, M. W. H. *Evolution of the Gas Industry.* London: Macmillan, 1980.

Pegrum, D. F. "Restructuring the Transport System." In *The Future of American Transportation*, edited by E. W. Williams Jr. Englewood Cliffs, NJ: Prentice-Hall, 1971.

Persson, T., and G. Tabellini. *Political Economics: Explaining Economic Policy.* Cambridge MA: MIT Press, 2000.

Phillips, C. F., Jr. *The Regulation of Public Utilities.* Arlington, VA: Public Utilities Reports, 1993.

Pierce, R. J. "Reconstituting the Natural Gas Industry, from the Wellhead to the Burnertip." *Energy Law Journal* 9, no. 1 (1988): 1–57.

Platts EU Energy. "Long-Term Contracts: A Legal Quagmire," no. 174 (Jan. 11, 2008), McGraw Hill, 8–9.

Posner, R. A. "The New Institutional Economics Meets Law and Economics." *Journal of Institutional and Theoretical Economics* 149, no. 1 (1993): 73–87.

Pusey, M. J. *Charles Evans Hughes.* New York: Macmillan, 1952.

Read, H. J. *Defending the Public: Milo R. Maltbie and Utility Regulation in New York.* Pittsburgh: Dorrance Publishing, 1998.

Reinhart, C. M., and M. A. Savastano. "The Realities of Modern Hyperinflation." *Finance and Development* 40 (June 2003): 20–23.

Report of the Commissioner of Corporations on the Transportation of Petro-leum (Garfield report). Washington, DC: US Government Printing Office, May 2, 1906.

Rostow, E. V. *A National Policy for the Oil Industry.* New Haven, CT: Yale University Press, 1948.

Rostow, E. V., and A. S. Sachs. "Entry into the Oil Refining Business: Vertical Integration Re-examined." *Yale Law Journal* 61 (1952): 856–914.

Samuelson, P. A. *Foundations of Economic Analysis.* Cambridge, MA: Harvard University Press, 1947.

Samuelson, P. A., and W. D. Nordhaus. *Economics.* 12th ed. New York: McGraw Hill, 1985.

Sanders, M. E. *The Regulation of Natural Gas: Policy and Politics, 1938–1978.* Philadelphia: Temple University Press, 1981.

Saussier, S. "Transaction Costs and Contractual Incompleteness: The Case of Électricité de France." *Journal of Economic Behavior* 42 (2000): 189–206.

Shaikh, Hafees, Manuel A. Abdala, et al. "Argentina Privatization Program: A Review of Five Cases." Case Study 3: Gas del Estado. Washington, DC: World Bank, 1995.

Sharkey, W. W. *The Theory of Natural Monopoly.* Cambridge: Cambridge University Press, 1982.

Simon, W. E. *A Time for Truth.* New York: McGraw Hill, 1978.

Smith, A. *The Wealth of Nations.* New York: Modern Library, 1937.

Spiller, P. T., and S. Liao. "Buy, Lobby or Sue: Interest Groups' Participation in Policy Making: A Selective Survey." In *New Institutional Economics—A Guidebook*, edited by E. Brousseau and J.-M. Glachant, 307–27. Cambridge: Cambridge University Press, 2008.

Splawn, W. M. W. *Report on Pipe Lines* (in two parts). H.R. Rep. No. 2192, 72nd Cong., 2nd Sess. Washington, DC: US Government Printing Office, 1933.

Stark, J. "The Wisconsin Idea: The University's Service to the State." Madison, Legislative Reference Bureau, 1995, p. 17. Repr. in *Wisconsin Blue Book, 1995–1996.* Madison, State Printing Office, 1995.

Stigler, G. J. "The Law and Economics of Public Policy: A Plea to Scholars." *Journal of Legal Studies* 1 (1972): 1–12.

Tadelis, S. "Complexity, Flexibility, and the Make-or-Buy Decision." *American Economic Review* 92, no. 2 (May 2002): 433–37.

Tarbell, I. *The History of the Standard Oil Company.* New York: McClure, Phillips, and Co., 1904.

Trapmann, W., and J. Todaro. "Natural Gas Residential Pricing Developments during the 1996–97 Winter." *Natural Gas Monthly* (US Energy Information Administration), Aug. 1997.

Trebing, H. M. "Martin G. Glaeser." In *Pioneers of Industrial Organization*, edited by H. W. de Jong and W. G. Shepherd, 190–93. Cheltenham, UK: Edward Elgar, 2007.

Troxel, E. *Economics of Public Utilities.* New York: Rinehart and Company, 1947.

———. "Long-Distance Natural Gas Pipe Lines." *Journal of Land and Public Utility Economics* (Nov. 1936): 344–54.

———. "II. Regulation of Interstate Movements of Natural Gas." *Journal of Land and Public Utility Economics* (Feb. 1937): 20–30.

———. "III. Some Problems in State Regulation of Natural Gas Utilities." *Journal of Land and Public Utility Economics* (Feb. 1937): 188–203.

Tussing, A. R., and C. C. Barlow. *The Natural Gas Industry: Evolution, Structure and Economics.* Cambridge, MA: Ballinger, 1984.

Tussing, A. R., and B. Tippee. *The Natural Gas Industry: Evolution, Structure and Economics.* 2nd ed. Tulsa, OK: PennWell Books, 1995.

UK Competition Commission, Office of Fair Trade. "Gas and British Gas PLC: Reports under the Gas and Fair Trading Act." London: Monopolies and Mergers Commission, 1992.

US Department of Justice. *Competition in the Oil Pipeline Industry: A Preliminary Report.* Washington, DC: Antitrust Division, Department of Justice, May 1984.

US Department of Justice. *Oil Pipeline Deregulation.* Washington, DC: Antitrust Division, Department of Justice, May 1986.

Vickers, J. S., and G. K. Yarrow. *Privatization: An Economic Analysis.* Cambridge, MA: MIT Press, 1988.

Weiss, L. O. "The Field of Industrial Organization at Wisconsin." In *Economists at Wisconsin*, edited by R. J. Lampman, 219–29. Madison: Board of Regents of the University of Wisconsin System, 1993.

Williamson, O. E. "Comparative Economic Organization: The Analysis of Discrete Structural Alternatives." *Administration Science Quarterly* 36, no. 2 (1991): 269–96.

———. "Credible Commitments: Using Hostages to Support Exchange." *American Economic Review* 83, no. 4 (Sept. 1983): 519–40.

———. *The Economic Institutions of Capitalism.* New York: Free Press, 1985.

———. "Franchise Bidding for Natural Monopolies—in General and with Respect to CATV." *Bell Journal of Economics* 7, no. 1 (Spring 1976): 73–104.

———. *The Mechanisms of Governance.* New York: Oxford University Press, 1996.

———. "The New Institutional Economics: Taking Stock, Looking Ahead." *Journal of Economic Literature* 38 (Sept. 2000): 595–613.

———. "Transaction Cost Economics." In *Handbook of New Institutional Economics*, edited by C. Ménard and M. M. Shirley, 41–65. Dordrecht, Netherlands: Springer, 2005.

———. "Transaction-Cost Economics: The Governance of Contractual Relations." *Journal of Law and Economics* 22, no. 2 (1979): 233–61.

———. "Transaction Cost Economics: The Natural Progression." *American Economic Review* 100, no. 3 (June 2010): 673–90.

Winston, C. "Lessons from the U.S. Transport Deregulation Experience for Privatization." Discussion paper no. 2009-10, OECD/ITF Joint Transport Research Center, Paris, 2009.

Wolbert, G. S., Jr. *American Pipe Lines: Their Industrial Structure, Economic Status and Legal Implications.* Norman: University of Oklahoma Press, 1951.
——. *U.S. Oil Pipe Lines.* Washington, DC: American Petroleum Institute, 1979.
Youngberg, J. C. *Natural Gas, America's Fastest Growing Industry.* San Francisco: Schwabacher-Frey, 1930.

Index